MARKETING BY CONTINGENCY IN THE TIME OF COVID-19

Overcoming Business Crises and Meeting Marketing Challenges

MARKETING BY CONTINGENCY IN THE TIME OF COVID-19

Overcoming Business Crises and Meeting Marketing Challenges

Edited by

Alicia de la Peña, PhD

Juan Bernardo Amezcua Nuñez, PhD

AAP | APPLE ACADEMIC PRESS

First edition published 2023

Apple Academic Press Inc.
1265 Goldenrod Circle, NE,
Palm Bay, FL 32905 USA

760 Laurentian Drive, Unit 19,
Burlington, ON L7N 0A4, CANADA

CRC Press
6000 Broken Sound Parkway NW,
Suite 300, Boca Raton, FL 33487-2742 USA

4 Park Square, Milton Park,
Abingdon, Oxon, OX14 4RN UK

Library and Archives Canada Cataloguing in Publication

CIP data on file with Canada Library and Archives

Library of Congress Cataloging-in-Publication Data

CIP data on file with US Library of Congress

ISBN: 978-1-77491-105-1 (hbk)
ISBN: 978-1-77491-106-8 (pbk)
ISBN: 978-1-00330-069-4 (ebk)

About the Editors

Alicia de la Peña, PhD
Full-Time Professor, School of Marketing,
Universidad Autónoma de Coahuila, Mexico

Alicia de la Peña, PhD, is a Full-Time Professor in the School of Marketing at Universidad Autónoma de Coahuila, México. Her research has been funded by the Mexican Secretary of Education. Her work has been published in the International Journal of Consumer Studies, the Journal of Nonprofit and Public Sector Marketing, and the International Journal of the Built Environment and Asset Management, and she has also published book chapters and several books, including De la Peña, Amezcua, Saucedo-Soto, and Hernández Bonilla (Eds.) *Estrategias de Consumo en un Entorno Dinámico y Cambiante* – 2019. Mexico: EMI – Autonomous University of Coahuila; Amezcua, De la Peña, Saucedo-Soto, and Hernández Bonilla (Eds.) *Tecnologías Digitales: un nuevo espacio público* – 2017. Mexico: INCUNÁBULA- Autonomous University of Coahuila; and (iii) De la Peña and Amezcua. *Responsabilidad Social Empresarial: Casos y estrategias de éxito en Mexico 2012*. Mexico: Plaza and Valdez – Autonomous University of Coahuila.

Dr. de la Peña is also a consultant in digital marketing strategies. She is a member of the National System of Researchers in Mexico (SNI) and the American Marketing Association (Mexico City Chapter). Dr. de la Peña's research interests include consumer behavior, consumer well-being, consumer health, and the use of social networks.

Dr. de la Peña earned her PhD in Business Administration at the EGADE Business School within the marketing/responsible consumer behavior research track. She earned a master's in Business Administration from Universidad Iberoamericana (2001) and graduated with honors from Tecnologico de Monterrey completing a BA in Marketing.

Juan Bernardo Amezcua Nuñez, PhD
Full-Time Professor, School of Marketing,
Universidad Autónoma de Coahuila, Mexico

Juan Bernardo Amezcua Nuñez, PhD, is a Full-Time Professor in the School of Marketing at Universidad Autónoma de Coahuila, México. He is also a business consultant in branding and market research and a member of the National System of Researchers in Mexico (SNI). He also contributes as Associate Editor to the international academic journal *Ciencias Administrativas*, Teoria y Praxis. He received the Mexico ANUIES National Award 2018 for innovation in teaching practice, intended to recognize teachers who throughout their professional career have been distinguished by performance innovation of high academic quality. He also has been given a best teacher award by Tecnologico de Monterrey in 2018. His research interest focuses on responsible consumer behavior and related topics, such as corporate/university social responsibility, social entrepreneurships, and the sharing economy, about which he has published more than 60 investigations, including 12 papers in international indexed journals and 16 book chapters.

His previously published books are: (i) De la Peña, Amezcua, Saucedo-Soto, and Hernández Bonilla (Eds.) Estrategias de Consumo en un Entorno Dinámico y Cambiante – 2019. Mexico: EMI – Autonomous University of Coahuila; (ii) Amezcua, De la Peña, Saucedo-Soto, and Hernández Bonilla (Eds.) Tecnologías Digitales: un nuevo espacio público – 2017. Mexico: INCUNÁBULA- Autonomous University of Coahuila; and (iii) De la Peña and Amezcua. Responsabilidad Social Empresarial: Casos y estrategias de éxito en Mexico – 2012. Mexico: Plaza and Valdez – Autonomous University of Coahuila.

Dr. Amezcua earned his PhD in Business Administration from EGADE Business School within the marketing/responsible consumer behavior research track. He holds a master's in Business Administration from Universidad Iberoamericana (2001) and graduated with honors from Tecnologico de Monterrey, completing a BA in Marketing.

Contents

Contributors

Juan Bernardo Amezcua Nuñez
School of Marketing, Universidad Autónoma de Coahuila, Mexico,
Tel.: +52-844-4129109, E-mail: bamezcuan@gmail.com

Cecilia Isabel Calderón-Valencia
Master of Management and Marketing, UPAEP University, Calle 17 Sur 901, Barrio de Santiago, Puebla – 72410, Mexico, Tel.: +(52) 222-229-9400, E-mail: dra.chechi@gmail.com

Gloria Camacho
Marketing and Analysis Department (Classrooms 6, Third Floor), Tecnologico de Monterrey, Campus Monterrey, Av. Eugenio Garza Sada – 2501 Sur, Tecnológico, Monterrey – 64849, N.L., Mexico, Tel.: +51-83-58-20-00, E-mail: Camacho.Gloria@tec.mx

Juan Ignacio Gonzalez Espinosa
Engineering Management Department at University of Monterey, Mexico,
Tel.: +52(81) 8215-1000; Ext. 1756, E-mail: jigonzale040@gmail.com

Eva M. Guerra-Leal
EGADE Business School, Monterrey, Eugenio Garza Lagüera and Rufino Tamayo, East Valley, San Pedro Garza García, Nuevo León – 66269, Mexico,
Tel.: +52-(81)-8625-6000, E-mail: guerra_eva@tec.mx

Ena Yuritze Barón López
Professor of the Department of Marketing of the Central University, Cra 9 No 60–69 – Chapinero, Bogotá, Colombia, Tel.: +57-3213214332, E-mail: ebaronl@ucentral.edu.co

Isabel Morteo
Faculty of Tourism and Marketing, Autonomous University of Baja California, Calzada Universidad – 14418, Otay Mesa Delegation, Tijuana – 22390, Mexico,
Tel.: +52-664-979-7578, E-mail: imorteo@uabc.edu.mx

Flor Morton
Business School, University of Monterrey, Administration Department,
Av. Ignacio Morones Prieto – 4500, Poniente – 66238, San Pedro Garza García, Mexico,
E-mail: flor.morton@udem.edu

Nancy B. Ortiz
Marketing and Analytics Department, Tecnologico de Monterrey, Monterrey, Mexico,
Tel.: +52-(81)-1290-4778, E-mail: nortiz@tec.mx

Alicia de la Peña, PhD
School of Marketing at Universidad Autónoma de Coahuila, México,
Tel.: +52-844-4129109, E-mail: aliciadelapenadeleon@uadec.edu.mx

Claudia Quintanilla
EGADE Business School Monterrey, Eugenio Garza Lagüera and Rufino Tamayo, East Valley, San Pedro Garza García, Nuevo León – 66269, Mexico,
Tel.: 86256010, E-mail: Claudia.quintanilla@tec.mx

Raúl Ruiz
Marketing and Analytics Department, Tecnologico de Monterrey, Monterrey, Mexico,
Tel.: +52-(81)-2724-6037, E-mail: rjruiz@tec.mx

Ana Ruiz-Medina
School of Business, Tecnologico de Monterrey, Monterey Campus,
Av. Eugenio Garza Sada – 2501 Sur, Tecnológico, Monterrey – 64849, N.L., Mexico,
Tel.: +52-1-818-396-1466, E-mail anatereruiz@gmail.com

Teresa Treviño
Business School, University of Monterey, Administration Department,
Av. Ignacio Morones Prieto – 4500 Pte., San Pedro Garza García, N.L. – 66238, Mexico,
Tel.: 8114135808, E-mail: teresa.trevinob@udem.edu

Ana Zavala-Parrales
Business School, Universidad Americana del Noreste, Saltillo Blvd Jose Musa de León #944, Coahuila, Mexico, Zip Code 25204
Tel.: +52-1-844-3627188, E-mail: azavala@uane.edu.mx

Abbreviations

AI	artificial intelligence
Apps	applications used on mobile devices
ARES	Coronavirus Aid, Relief, and Rconomic Security Act
CBE	customer brand engagement
CBK	company's brand knowledge
CCSRC	consumer-centric socially responsible consumption
CKO	Chief Knowledge Officer
COVID-19	coronavirus pandemic
CSR	corporate social responsibility
DIY	do it yourself
eNPS	employee net promoter score
FAQS	frequently asked questions
FLEs	front line employees
GA	grassroots associations
GIS	Grupo Industrial Saltillo
HR	human resources
HRM	human resource management
IM	internal marketing
MPP	management policies and practices
NPO	non-profit organization
OC	organizational commitment
OCB	organizational citizenship behavior
OECD	organization for economic co-operation and development
POS	perceived organizational support
QR	quick response code
RVB	resources based view
VUCA	volatile, uncertain, complex, and ambiguous
WHO	World Health Organization
WOM	word-of-mouth
Y2K	year 2000

Acknowledgments

We would like to extend our appreciation to Dr. Cristóbal N. Aguilar, Director of Research and Graduate Studies of the Universidad Autónoma de Coahuila (UAdeC), and his team for encouraging us to start this editorial process.

We also would like to extend our recognition to the 12 chapter authors, who made great contribution to this book as well as to all the reviewers who provided knowledgeable feedback in a rigorous double-blind peer-review process. Your work makes this book a high-quality contribution to knowledge advancement in marketing strategies during a worldwide crisis.

Preface

COVID-19 came with no warning, leading organizations to implement market strategies with no time to prepare. Almost no organization was prepared to face the challenges imposed by this global health contingency that suddenly turned the business environment upside down. Only a small group of resilient organizations responded at the first stage of the contingency, while the majority of them waited to see consumer reactions to the risky marketing strategies implemented by their competitors.

Even public policies came late to show the right path for organizations to act and for consumers to respond. Marketing strategies were diverse and not always with satisfactory results. The 12 chapters in this book analyze successful and failed marketing strategies implemented by renowned and emerging brands. The authors took the challenge to identify those elements that were critical for strengthening brand value, consumer satisfaction, and loyalty in times of crisis and uncertainty. The ideas, conclusions, and recommendations in this book contribute to the literature on brand strategy and consumer behavior during health or any other contingencies faced by organizations.

In sum, this book looks to assess how organizations learned to be resilient and act upon unprecedented market situations, using creativity and innovation to present consumers with prompt solutions to cover their most basic needs and alleviate their fears.

The book, *Marketing by Contingency in the Time of COVID-19: Overcoming Business Crises and Meeting Marketing Challenges*, provides theoretical foundation, conceptual framework, and practical application of marketing strategies.

In order to provide readers with a valuable academic resource, at the end of each chapter the authors have included—and defined—those key terms that are more relevant and that were used to explain certain marketing and business phenomena.

We are sure that the 12 chapters will be useful for academic researchers, business managers, teachers, and students who are interested in understanding how the COVID-19 pandemic affected consumers' life, businesses' marketing strategies, and brand building.

Introduction

Marketing history has shown us that firms and organizations that closely pay attention to their external environment, and to what their stakeholders need, want, or demand at a certain moment, are the ones that successfully thrive and remain relevant over time.

We just need to look around, and we will find companies like General Electric, Boeing, Coca Cola, Kelloggs, or UPS that have been in the market for over a century selling goods and services that perfectly deliver what the 21st century customer needs. Such firms have successfully faced what the World Economic Forum calls the Second and Third Industrial Revolutions, as well as wars, climate changes, and cultural shifts, among other challenges that have required that CEOs, managers, and marketing executives act with agility, flexibility, and creativity.

And just when CEOs and public leaders were facing the Fourth Industrial Revolution (i.e., the digitization era), an unexpected new challenge arrived: the COVID-19 pandemic. A small and invisible but nevertheless powerful virus that made public officers, entrepreneurs, managers, and citizens to rethink the way they govern, work or live.

This volume, *Marketing by Contingency in the Time of COVID-19: Overcoming Business Crises and Meeting Marketing Challenges,* contains 12 chapters written by academics and researchers, experts in the fields of business and marketing, who during the first months of the COVID-19 pandemic had the opportunity to observe and investigate how the world-wide health crisis rewrote the business rules, forcing managers and creatives to redesign products, services, delivery systems and even advertising and promotional strategies to suit what has been called the "new normal."

The VUCA (volatile, uncertain, complex, and ambiguous) environment faced by governments, firms, and non-government organizations left leaders and managers with no other choice than to modify their current strategies and quickly innovate to attend to the challenges created by the pandemic. Some of those strategies were investigated by the authors of this volume.

This book contains both academic and market research, discussion of theoretical foundations, business cases and insights on managerial implications. In the different chapters, the authors address the challenges faced by firms' managers, not-for-profit leaders, and of course, the consumers. In order to help readers to better understand each chapter, the authors provided a list of the most important keywords and a definition of each word. The keywords are presented at the end of each chapter.

The present volume addresses the study of consumer decision making while under specific emotional states such as fear, anxiety or sadness. The research methods employed by the authors will allow readers to gain a deeper insight into consumers' decision processes while living under lockdown measures. It is also important to mention that some of the chapters move beyond consumer research and study firms and their organizational strategies. Specifically, the authors mention the importance of using technology and digital tools to comply with the new government policies as well with the consumers' needs.

The book begins with a look at the importance of empathetic brands. In the first chapter, Amezcua examines how well-known brands of digital devices missed the opportunity to show a more human side and provide their current and future customers with empathy and understanding. In the second chapter, based on a qualitative research, Quintanilla presents a review of customers' journey during the pandemic and their engagement with those brands that demonstrated their concern for consumers' health and well-being.

Empathy and concern were not the only attributes sought both by managers and consumers. Resilience, creativity, and flexibility became important factors while dealing with the lockdown and social distance imposed by public authorities to prevent the spread of the virus. De la Peña provides some insights on how small firms adapted their strategies to the new needs of individuals who, forced to work, study, and exercise at home, suddenly changed their consumption patterns.

E-commerce, curbside pick-up, and intensive use of social media were some of the tools that firms and organizations had to use to help consumers solve their problems and cater to their whims. While Morteo sheds light on the importance of using well-known influencers as strategic partners and their role on delivering relevant and accurate information to their publics, both Barón and Camacho present an analysis of the omnichannel strategies used at the retail level; and, Calderón-Valencia describes the response given by nonprofit organizations during the pandemic.

Gonzalez-Espinosa's chapter provides us with a deeper understanding on the use of big data analytics to develop predictive models in order to be able to make better decisions and avoid pitfalls during a crisis.

The effects of emotions on consumers' decisions are also considered in the book. For most readers it is well known that fear and anxiety were present almost since the beginning of the pandemic. The uncertainty of the disease and its consequences moved individuals to change their consumption patterns, as Treviño (in Chapter 5) Zavala-Parrales and Ruiz-Medina (Chapter 8) explore. Cleaning supplies and food were not the only items bought by consumers. Credible information became a staple, and trust on the source of the news soon became an important factor. As Morton depicts in her chapter, during the lockdown, consumers needed to trust firms, and corporate social responsibility (CSR) became an important pillar for most companies.

Facing a crisis like the COVID-19 would be impossible without the involvement of each one of the internal stakeholders. To give a better understanding of the internal marketing (IM) strategies used by firms to motivate and take care of the employees and collaborators, Guerra-Leal, Ortiz, and Ruiz conducted an interesting study, that measured organizational commitment (OC) and employee net promoter score (eNPS), an important marketing tool that allow us to measure employees' loyalty and commitment.

In sum, the COVID-19 pandemic left us with a deeper knowledge and understanding of consumers' reactions and behaviors during stressful situations as well as the managerial responses taken under pressure and sometimes with limited resources.

I thank the authors for their valuable contributions to this book and for their interest to enrich the marketing world by providing students, academics, and practitioners with business and marketing cases. I am sure each chapter and each case will become an important tool in different marketing and business courses.

—Alicia de La Peña

CHAPTER 1

Alexa, Show Some Empathy: Personified Brands Failing to Engage Customers During COVID-19

JUAN BERNARDO AMEZCUA NUÑEZ

School of Marketing at Universidad Autónoma de Coahuila, México, E-mail: bamezcuan@gmail.com

ABSTRACT

The novel coronavirus pandemic (COVID-19) and its consequent mobility restrictions have led consumers to a state of uncertainty and astonishment never seen before. Brands experienced a similar situation, hesitating about how and when to act in response to the pandemic. Some brands opted for a cautious strategy, waiting to see how the pandemic developed. Some others stood up rapidly to express their understanding of the situation and accompany consumers in the process. Cautious brands, however, could have been perceived as selfish companies looking after their own interests and not caring for people in distress. Especially those brands that benefited from the lockdown, since their service demand increased notoriously, should have communicated with users from day one. In addition, personified brands may be expected to stay closer to consumers since, as posited by identification theories, they use anthropomorphic elements-resembling human traits – to engage them. Most of the personified brands missed an opportunity to early engage with users by showing some empathy at the beginning of the pandemic to increase brand credibility and loyalty. Brand theories talk about failures and transgressions. Most personified

Marketing by Contingency in the Time of COVID-19: Overcoming Business Crises and Meeting Marketing Challenges. Alicia De La Peña, PhD & Juan Bernardo Amezcua Nunez, PhD (Eds.)

brands seem guilty of omission at some point of time by doing nothing about pandemic until there was an official position from World Health Organization (WHO). This theoretical paper based on a throughout literature review and evaluation of the Alexa brand performance during the COVID-19 pandemic adds to brand transgression literature by assessing the moderating effect of brand personification.

1.1 INTRODUCTION

Like many other brands, Alexa uses anthropomorphic elements to embody human traits, which facilitates bonding with users. Personified brands might have found it easier than ever to bond to consumers through affect and emotion during the pandemic. And most of them fail to do so.

The year 2020 has been challenging for brands and their relationship with consumers. COVID-19 has generated a great deal of uncertainty both for companies and consumers.

On the one hand, consumer behavior has adapted to changes in context derived from restricted mobility and confinement. Consumers flooded with information, not always from trustable sources, felt insecure about how to act responsibly with themselves and others. In the first stages of the pandemic, information from global health authorities was scarce and changed from one day to another.

Brands experienced a similar situation, considering that the coronavirus crisis was novel and no person or organization in the world had a clear and definitive idea on how to prevent it or treat it. In this chaotic context, personified brands had an opportunity to become a lifeguard figure by finding ways to bring confidence to consumers while adding value to the brand.

Especially at the beginning of this pandemic, brands seemed unsure about whether to communicate with consumers, how to do it and what to tell about it. They couldn't find a wise position from which to engage with their customers' feelings and emotions. Many of them were reluctant to refer to coronavirus, as they did not want to portray themselves as beneficiaries of the crises. They did not want to look like opportunists profiting from a health crisis. However, personified brands through their human resemble could have added value to people's lives and found a touchpoint to engage consumers. Those who didn't relate to coronavirus missed an incredible opportunity to show their customers that they care and share their thoughts about the pandemic.

This chapter draws on brand anthropomorphism and identification and social theories to assess consumers' responses to brand omissions. Omission neglect (Kardes, 1988) is a phenomenon related to information processing characterized by a lack of sensitivity to missing or unknown information. In a context of anthropomorphized brands (also known as personified brands) where individuals commonly grant human characteristics, motivations, and behaviors to non-human entities (e.g., brands), consumers are prone to perceive a brand as an intentional agent capable of interacting with them and establishing relationships that resemble those between actual people. Brand anthropomorphism may drive purchases based on a positive relationship with customers. However, when brands by accident, intentionally, or by omission breaks the assumed values of the relationship (e.g., omitting or hiding information), consumers may feel betrayed and disappointed.

Research literature has studied phenomena like brand transgression, service failure, and recovery, and brand-harm crisis using complementary theories, concepts, and methods. They all contribute to know best business practices to deal with consumer discomfort. However, little has been said about anthropomorphized or personified brand omissions and their effect on brand credibility and brand loyalty.

1.2 BRANDS THAT BENEFITED FROM THE PANDEMIC

Lockdown and physical distancing measures globally imposed contributed to increase economic activity towards online platforms. Global statistics about online platform use vary across activity areas. Activity areas, which require physical proximity (e.g., transport, accommodation, and restaurant services, social activities) declined around 70%. In contrast, those activities not requiring physical proximity for product and service delivery (e.g., digital entertainment, professional services, electronic payments, online market places, food delivery) increased online platform use by about 20% (OECD, 2020).

Demand for products and services for primary needs, such as medical items, food, education, entertainment, and medical services online, clearly increased (Nielsen, 2020). Nielsen World panel service reports double-digit growth data for those categories. However, despite their rapid growth at the beginning of the pandemic, medical, and food items are expected a following period of enactment in which the products in stock are consumed.

The spread of the virus and the stress lived at home because of confinement affected people's usual spending and consuming behavior. Consumers change their daily habits and, consequently, the products or services acquired. Even though lockdown is being rigorously carried out by only 4% of consumers in the world, and they nowadays prefer to buy local (e.g., in the corner store and what is made locally); 4 out of 10 have started to purchase all kind of products by phone or online. They are trying to eat healthier, exercise, and carry out a better planning of purchases as they would look for more information from choice brands (AMVO, 2020; Entrepreneur, 2020). For instance, 5 out of 10 companies in Mexico are doubling their growth in online channels, and 2 out of 10 registered increases greater than 300% in the volume of online sales (AMVO, 2020). With fewer people exiting the home, sales of items related to offline activities decreased. Despite the lower consumption, specialists agree that consumers maintained their preferences for their favorite brands. Seven out of 10 buy the same ones they used before the crisis, with 20% saying they are willing to try new ones. However, consumers appealed to brand empathy in recognition of life challenges imposed by confinement (Garcia-Fuentes, 2020).

1.3 LIFE UNDER CONFINEMENT AND CHANGES IN CONSUMER BEHAVIOR

In some countries, by law, people were not allowed to leave their houses but to do basic activities to keep the household running. All over the world, schools, and universities were closed at times, and homeschooling became the new normal. Similarly, home office was one of the winners of this crisis, and most public events were canceled. Even without strict lockdown measures, people used to spend much more time at home. Even in countries that have not been severely affected, many people isolated themselves voluntarily. Air travel, public transportation, or large gatherings of people were avoided. This inevitably meant that people spend much less time socializing with others and spend much more time at home, alone, or with their family. Isolation brought some benefits to consumers whose life slow down, and they could spend more time with their family and carry out those personal activities that had been postponed due to lack of time. Consequently, they started to use more services provided through the Internet (e.g., music, and video streaming services, online shopping, home deliveries, etc.). Even though, as lock-down prolonged, people had to get back to their daily activities outside their homes. For instance, according to AMVO (2020), some kept going out almost as they did before in Mexico. Still, being cautious (18%), the majority went out to see friends and family or went to work but limit any other exits (48%). Part of the population kept relatively isolated, going out only for food or in case of medical emergencies but limiting contact with other people (29%). Finally, only a minority remained in total quarantine, nobody was entering or leaving their houses and what they needed was home delivered (2%). For those living in total, forced, and prolonged isolation caused anxiety and a sense of loss of control over their lives.

The continuing spread of the virus and the stress lived at home affected people's lives and their spending and consumption behavior. Consumers changed their daily habits and, consequently, the products or services acquired. People started consuming more multimedia content and more games. They began to participate in other activities around the house. Since they could no longer go to the gym, they began to exercise indoors. People spent much more time cleaning and disinfecting. And as people avoided eating out, they began to cook more often or started using delivery apps (applications used on mobile devices). As people were trapped at home, they avoided going out (e.g., partying, clubbing) and started participating in online parties and tuning in to video platforms like TikTok and other online socialization spaces. Fashion being a form of expression of one's identity, changed more radically to online channels. Consumers showed some signs in favor of sustainability and the detriment of fast fashion (Entrepreneur, 2020).

Going back to the new normal has led firms to adjust the way they do business offline. Some companies reopen their establishments under new safety norms to ensure their employees' and customers' health: hygiene measures, packaging, cleaning, sanitization, touch-less payment methods, etc. (Regadera-Gonzalez, 2020). New service and health standards have contributed to a promising economic' rebound.

1.3.1 SENTIMENTS AND SOCIAL COMMUNICATION

With prolonged lockdown, people experimented some negative sentiments. They felt anguish about money issues (40%), and uncertainty about what will the future bring (40%). Not knowing what family/friends nearby have been infected (34%) or the risk of getting infected (32%). The decrease in the quality of life due to not being able to go out (30%). Loss of income because of losing job (22%). Insecurity, robbery, and looting (19%). Not being able to travel or move to other cities/countries (18%) (AMVO, 2020).

To diminish anxiety, people started to try new things at home to keep busy and clear their minds. After six months in isolation, consumers increased their activity in social media as a way to communicate with others and express their feelings. For instance, following brands they like through social networks increased 23% of all actions taken by online purchasers. Subscriptions to brand communication channels they like and sharing what they shopped online through social networks slightly grow during pandemic (to 14% and 7%, respectively). They were positing their expectations on the brands their trust to help them lead with confinement. However, writing reviews about their experience with the product or service purchased online dropped from 20% to 13% (AMVO, 2020). Online statistics confirmed that facilitating content marketing became more relevant for digital users in lockdown. The use of emotions became relevant to consumers that needed empathetic listeners and brands.

Nevertheless, brands showed a poor reaction. According to AMVO (2020), publications made by retail brands in Mexico on social networks (Twitter, Facebook, and Instagram) after the pandemic outbreak remained almost the same as before. It was until September 2020 that retail brands incremented their posts by 18% on Instagram, 5% on Facebook, and 3.1% on Twitter. One reason for this growth was back-to-school sales and promotions and did not directly mean a response to the pandemic. During the pandemic, brand communication pointed to Twitter as the principal place to share, reaching 49% of all online posting. Facebook held 46% and Instagram only 5% of all brand publications on social networks. Interactions with users showed a different picture, having Facebook as the leading platform where consumers interacted with brands (75%), followed by Instagram with 21% and Twitter with just 4% of total interactions (AMVO, 2020).

For categories such as department stores and supermarkets, Facebook was the social channel with the highest engagement rate. Consumers' interactions with retail brands did not change after the pandemic outbreak until September 2020 where Twitter and Facebook showed a slight increase in their communication activity. Instagram did not show any change during the pandemic (AMVO, 2020). Big retailers facilitated and encouraged online shopping by offering free shipping, purchases without a membership, trial periods, or the inclusion of holiday promotions that happened in the middle of the pandemic.

More importantly, talking about consumer sentiment toward online shopping during pandemic increased positively. Share of sentiment change is shown in Figure 1.1.

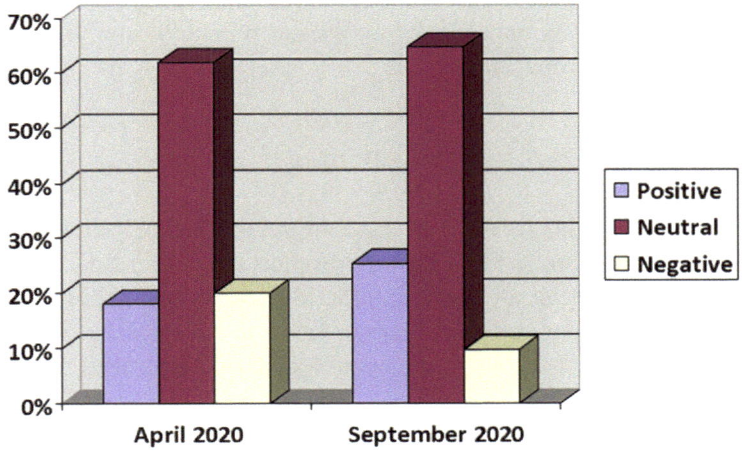

FIGURE 1.1 Sentiment share for online shopping on social networks.

Source: Own with data from: AMVO (2020).

Positive sentiment about online shopping increased from 18% to 25.4% and negative sentiment decreased from 20% to 9.8%. Main topics in social networks focus on: free shipping and holiday promotions, but also on fraud in consumer deliveries and other services. However, at the beginning of the pandemic, negative sentiment almost doubled the September numbers, indicating that scarce information and overall brand standard marketing strategies (avoiding pandemic topics) did not offer confidence to consumers.

Affective brand communication, as has been demonstrated, significantly affects consumer behavior. The use of emotions in messages addressing consumers can lead to a competitive advantage and increased brand equity (Ang et al., 2018). In addition, emotional digital marketing help to build consumer engagement with the brand, considered as a psychological state of mind produced by interactive behavior (Syrdal and Briggs, 2018). In this context, emotional branding may be defined as the process of building a relationship between consumers and brands through the generation of emotions. Establishing an emotional bonding with consumers allows brands to differentiate from the competition by reinforcing positive brand recognition. According to some studies, 90% of buying decisions are made subconsciously, individuals process visuals with emotional content 600,000 times faster than regular text, and 50% of brand experience is based on emotion. Emotional connections eventually lead to brand loyalty, translating into greater retention rates and customer lifetime value (Karnes, 2020).

1.3.2 BRAND CREDIBILITY AND BRAND LOYALTY

Brand credibility has its roots in the communication and interaction with customers. At the same time, brand presence and brand salience are two dimensions impacting a brand's interaction with consumers (Stabler and Fischer, 2020). Brand salience refers to the level of activation in long-term memory, meaning these are brands positioned on consumers' top of mind, and their position does not change frequently. Brand presence refers to its presence in consumer's short-term memory at a specific point in time. Consequently, brand presence fluctuates over time due to influencing factors such as advertising, online presence, and word of mouth (both off and online). Brand salience, in particular, has been demonstrated to lead consumers to trust the brand. The more brand salience, the bigger trust in the brand. Stabler and Fischer (2020) identifies relevance as an essential element to increase brand salience. Relevance refers to brand stories that are perceived to be of importance to consumers. Subjects or things that are culturally similar, or affect several markets or nations, are likely to be considered more relevant because they fit into the consumer's frame of reference. Brands communicating relevant issues to consumers are perceived as trustable.

Brands must be authentic in their stories to achieve relevance; otherwise, consumers perceive opportunism and a simulated interest. In this respect, the issue of purpose is fundamental to brand trust because a brand is conceived to generate value for its stakeholders; those who do not manage with purpose will be those who will not achieve the trust of their consumers (Garcia Fuentes, 2020). Consumers' loyalty is more likely to be developed toward credible authentic brands.

Brand loyalty has been traditionally conceptualized from a behavioral and attitudinal logic (Copeland, 1923; Day, 1969; Jacoby, 1971). Behavioral loyalty focuses on consumers buying a brand repeatedly. Attitudinal loyalty, on the other hand, refers to what consumers think of a brand. Oliver (1997) goes further to propose a multidimensional construct of brand loyalty expressed in different phases. To Oliver (1997), brand loyalty starts in a cognitive phase where consumers assess available information in relation to brand performance and other competing alternatives. Then, loyalty takes on an affective nature, where consumers express positive emotions toward the brand. The last stage is the action phase, where consumers voice their good intentions by purchasing the brand more than once.

In addition, brand loyalty has been successful in attracting engaged customers where consumption is based on identity and socialization (Belk and Tumbat, 2005); this is when brands become part of a personal experience. For Pine and Gilmore (1999) brand experience might be captured, among other possibilities, by educating consumers about the environment through a relationship that helps to connect customers with what is happening in the world. Brands usually create emotional communications in response to major events showing that they care for what happens globally that might impact consumers. Finally, beyond loyalty, brands help build identity narratives whose meanings help in the construction of customers' identities by aligning shared values (Obiegbu et al., 2020).

1.4 PERSONIFIED BRANDS AND BRAND SINS

At the beginning of this crisis, governments faced critical shortages of medical supplies, which led them to compel private firms to change production to meet high demand. Governments in different countries forced private non-medical companies (e.g., General Motors) to produce ventilators, face masks, and hand sanitizers, however, some other companies from different industries have voluntarily shifted their manufacturing to make health care supplies, among them automakers like Ford, Fiat Chrysler, Toyota, and Volkswagen, fashion brands like Chanel and LÓreal or beer and spirit makers like Anheuser-Busch and Diageo. Being its motivation pure altruism, or a strategic way to bolster reputation and avoid excessive regulation, these brands stepped up rapidly in the eyes of the public (Mafi-Kreft and Kreft, 2020).

Especially at the beginning of the pandemic, people needed reassurance from trusted sources. Since brands pursue establishing long and trusted relationships with consumers, it is frequently seen as a credible source that helps mitigate anxiety derived from a shortage of information. If brands fail to provide information about difficult events, it is perceived as a purposeful omission seeking to highlight their usual positive attributes and omitting those that might represent a potential risk.

When digital marketing falls short in providing sufficient information, it will not lead to brand equity or deliver any benefit to the brand (e.g., brand meaning, brand loyalty); on the contrary, it may destroy value, increase cost transactions, and have a negative impact on long-term benefits (Aswani et al., 2018). Short of information represents a brand transgression. In this context, brand transgression is defined as a form of brand failure which breaks an implicit or explicit rule and does not meet customer expectations (Magnusson et al., 2014). Brand transgression can include accidental (e.g., industrial accident) or intentional (e.g., low service performance) failures that positively affect the relationship with customers in the short and long term. Notably, brand transgression might involve interpersonal losses related to the brand's attitude and how it communicates with consumers. This kind of brand transgression is called relational (Bitner et al., 1990) because it affects the relationship with consumers. The bad news is that consumers' forgiveness may be less common when a relational transgression is presented because it is perceived as a brand omission that could have been avoided. In these cases, brands are more severely criticized because they don't seem to be interested in consumer welfare. Consequently, brand omission is characterized by the lack of sensibility toward unmentioned or unknown information (Kardes, 1988). However, consumers not always may identify the absence of information, and those who are not experts on a subject are less sensitive to brand information omission (Lopes, da Silva and da Costa Hernandez, 2013).

Consumers often consider brands as intentional agents and perceive their relationship as similar to that between people (Fournier, 1998).

Consequently, brands are perceived as human agents rather than inanimate objects, and the relationship with consumers becomes similar to those between people. Consumers expect that brands respect and honor their relationship with them. Before a brand transgression, empowered consumers respond unfavorably to personified brands because they experience resentment due to broken promises in the relationship. Brand humanity (as personified or anthropomorphized brands) is defined as a collection of human traits and the enactment of compassion, benevolence, generosity, kindness, and empathy (Krishen and Berezan, 2019). In the age of big data and analytics, communication channels are filled with interactions between machines and humans that implies some vulnerability to users due to a lack of humanity in automatic initiatives.

Especially with well-known brands, consumers expect an impeccable relationship. Despite the fact that their disadvantaged position in the market compared with top brands, less known brands are more likely to receive consumers' support and generosity based on their identification with each other. Consumers are prone to forgive brand transgressions more easily when the brand is not the market leader (Kim et al., 2014). Less known brands, although with fewer resources, are seen as more passionate and committed to consumers. Conversely, with top brands, who are seen as having high resources but sometimes low passion and determination, minor transgressions may be magnified, and customers are less willing to forgive. However, brand identification may serve as a mitigating attribute because it helps to anthropomorphize brands. Personified brands are seen by consumers as close, mindful, and sympathetic (Eppler and Waytz, 2010). The relation between a consumer and a brand may be seen as one with a friend. A fault committed by a friend is more easily forgiven than the one committed by a stranger. Finally, consumers who share a strong identification with a brand are less influenced by brand failures or negative brand publicity and even exhibit higher brand immunity to brand transgression effects (Kim et al., 2014).

1.5 METHODOLOGY

A literature review on brand transgressions and personified brands in international academic journals was conducted. Based on Identification and Social theory, all the information was assessed to integrate a framework that facilitates a more profound understanding of how consumers perceive and respond to transgressions when performed by a personified brand during a crisis (i.e., pandemics). In addition, an e-mailing campaign from Alexa (Amazon) was evaluated from March to October 2020 to identify if communication supports consumers in any way during the COVID-19 pandemic. The Amazon Alexa web page was reviewed to identify supporting resources to help users get over pandemics. Social networks were spotted for negative posting on Alexa during the same period.

1.6 RESULTS AND DISCUSSION

Some brands drew a lot of attention from consumers worldwide, when in the early days of the pandemic, they modified their logos to show their empathy. Brands like Coca-Cola, Audi, Olympic Games changed their logos by spacing their elements resembling social distance required by health authorities. Starbucks siren started wearing a safety mask, and KFC admitted its famous Finger-Lickin' Good slogan was not quite right amid pandemic and official hand-washing advice.

So they released new advertising images and packaging with the words blurred and pixelated. In Latin America, the number one online store Mercado Libre also acted rapidly. The classic handshake that stands out since 1999 in the logo of this firm was replaced temporarily for an elbow salute. Most of these brands have some kind of anthropomorphic features, making its message meaningful to consumers. Other personified brands like Alexa or Siri, even when their business grew with pandemics, showed a slow and cautious response.

Alexa is the virtual assistant developed by Amazon that works using voice control (i.e., voice recognition). Individuals may ask questions out loud, and their personal Alexa device will respond. When someone asks a question, the device record his/her voice sends it through the Internet to Amazon's Alexa Voice Services, where the recording is transformed into specific computer commands that enable the system to send back a relevant output. Alexa, as other competing voice assistants, is presented as female. It has a feminine name and offers a woman's voice as people seem to prefer it. A female voice is also less authoritarian than a female male voice, which helps to establish a closer and more trustable relationship. Even though Alexa female personality is also perceived as more supportive and more willing to help someone (Elks, 2019) did not show up in the pandemic early stages.

Amazon Alexa offers consumers to subscribe to the Alexa Developer Newsletter, where the latest news and features about Alexa capabilities are sent to your e-mail. Two e-mails are sent each week.

For this study, more than 60 "Alexa News" e-mails were assessed. The letter format comprises three sections: (i) header; (ii) what's new this week; and (iii) the most requested these days. Alexa e-mail header highlights a significant event happening every week. Then what's new section makes suggestions of new phrases that Alexa should understand and respond to. Finally, the most requested these days' section ranks phrases more used by consumers.

There was no space for the intended transgression analysis in the e-mailing campaign. Since none of the information was ever about COVID-19, lockdown, managing stress generated by confinement, or any other related issue. Either e-mail section did not mention anything about this subject and no question about pandemic ever appeared as a new Alexa's ability. Up to March 2020, if asked directly to the Alexa device to inform about the latest news for COVID-19 a vague and elusive response was received. It seemed that COVID-19 did not exist in the Alexa reality. It was until the end of April 2020 that some tech blogs reported that Alexa was now capable of answering about COVID-19 and even could help to review (not fully diagnose) if you present some COVID-19 symptoms. Alexa users could also ask about the daily cases of COVID-19 in your country and the rest of the world, as well as the latest news and measures that have been taken in the country. Alexa would respond with the most recent update, according to information and official numbers published by the corresponding government office (ConsumoTIC, 2020). Alexa was now capable of answering questions like:

- Alexa, how many cases of coronavirus are there in ____ (name the country)?
- Alexa, how many cases of coronavirus are there in the world?
- Alexa, tell me coronavirus news.
- Alexa, how can I prevent coronavirus?
- Alexa, what are the symptoms of the coronavirus?
- Alexa, when does the quarantine end?
- Alexa can also help people make a homemade mask. Just by saying.
- Alexa, how can I make a homemade mask?

I then decided to explore The Amazon Alexa web pages to explore what the brand had to communicate about the pandemic and how they help consumers cope with the pandemic.

In April 2020, in the help and attention section of the Amazon web page, some COVID Alexa resources were displayed and recommended to use Alexa to keep yourself productive, connected, and entertained.

Although Alexa reconsidered her interaction with consumer's over-time, in the first stage of the pandemic, she did not show any resources that would allow her to reach out to consumers to support and build brand trust. That could be taken as a brand transgression that could harm brand identification and loyalty, but consumers did not show any discomfort about it on social networks. No relevant negative publication was found either on Facebook or Instagram. On Twitter appeared some rebel accounts using the Alexa name with political hues (e.g., @Alexa, buy me a covid vaccine) unrelated to the Amazon assistant device. In fact, in September 2020, during the Amazon fall product launch, the firm CEO emphasized the value of a connected house for those stuck at home and promoted consumer interest in videoconferencing, security systems, and streaming services. However, some might see Alexa growing influence in people's lives as dangerous because one of the wealthiest firms in the world is behind her, and consumers are allowing them to collect a lot of personal data. For those consumers concerned about safety issues, it seems brand trust seems to be compromised.

1.7 CONCLUSION

This chapter contributes to the existing literature on consumer-brand emotional relationships by assessing consumer behavior in response to perceived transgressions by personified brands to which they relate based on identification.

The pandemic has largely generated anguish over money and uncertainty issues for not knowing what will happen at the end of 2020 or 2021. After almost a year of pandemic, several consumers began to resume their usual activities, where 38% go out to work but taking safety measures. Another 29% are pretty isolated; however, their outings are limited to buying food, medical emergencies, and restricting contact with other people. In general, since the COVID-19 pandemic started, the reasons for purchasing online have remained unchanged, as these focus on not leaving home, avoiding agglomerations, and opting for home delivery. Pandemic also brought out several problems for online shopping related to brand performance (e.g., delivery times, quality issues, safety delivery procedures). Confidence in the online purchase process showed a significant growth due to brand performance and its attention to the new consumer journey. Brands had to get creative and adapt to the situation in order to keep consumers loyal. Some have joined last-mile platforms such as Rappi or UberEats, while others decided to opt for click and collect. Lockers to collect orders have also become more common, and companies are building new logistics hubs to meet the demand for orders. Contactless payment methods are preferred, so the use of bank applications and QR (quick response code) payments have been a great option to replace payments with physical cards.

Along with brand performance, credibility, and trust generated by personified brand communication are a significant generator of value for a brand that seeks to engage consumers. Beyond brand performance, it is essential to bring brands and consumers closer thru functional, valuable, and relevant content that helps consumers out of the pandemic crisis and guides their decision-making. Brands must tell stories that make it clear to consumers that the brand's best interest is in helping them improve their lives and reduce or mitigate risk in the new normal (Garcia-Fuentes, 2020). Missing to fulfill consumer emotional needs may be perceived as a brand transgression.

The study of significant conflicting events occurring between brands and consumers is important in the development of consumer relationships. Brand transgression, understood as posited by Aaker et al. (2004), is an act of violation of implicit or explicit rules between brands and consumers that impacts performance and evaluation. Brand transgression has been studied from different angles, among which is anthropomorphism (as a personified brand). Brand behavior violating the norms (Sayin and Gürhan-Canli, 2015) or social codes (Lin and Sung, 2014) of the brand-consumer relationship may serve as defining moments that lead to negative emotional consequences. Consumer emotional responses to personified brand transgressions may vary, going from discrete (disgust, disappointment), inward-focused (e.g., regret, and shame), to outward-focused (e.g., anger, and rage) feelings that may trigger desires of avoidance, reconciliation, or revenge (Khamitov et al., 2020). Customers experience emotional drive, leading them to take a sort of action related to negative online messaging and buying behaviors. What determines the type of response is in part determined by both consumer and brand personality. Brand personality literature is based on the conceptualization that consumers are more attracted to brands that resemble their personality. Consumers identify themselves with brands so that adventurous consumers will appreciate spirited brands, and traditional consumers will identify themselves with brands that evoke traditions and stability. Thus, consumer and brand personality literature based on identification theory may help to understand why some brand omissions are perceived as more detrimental than others. One of the most spread brand personality models is the Big-Five (Goldberg, 1993). Among those traits, conscientiousness refers to trusted brands where easy-going consumers might easily tolerate brand failures. Similarly, Agreeableness is based on qualities such as affection and kindness that can be associated with positive emotions, and brands scoring high on this trait are less commonly target of consumer's negative emotions and behaviors.

However, some authors (e.g., Patterson and O'Malley, 2006) question the literal conceptualization of the human-like relationship between customers and brands. For them, the taken-for-granted idea of a mutually beneficial relationship with brands is an idealization. Sometimes customers remain loyal despite unsatisfactory brand actions, and occasionally loyal consumers may act against the brand (e.g., negative word of mouth) and at the same time still have to opt for this brand when buying.

Consumers seem to be insensitive to many types of omission because they are novices (i.e., they are not experts on the subject) and are more sensitive to the omitted information. Past studies suggest that consumers frequently make judgments and decisions based on incomplete or biased information. In times like this, the interaction between brands and consumers can trigger positive consequences, increasing the degree of loyalty and trust since a personified brand, by presenting itself as a crucial solution, could satisfy the empathy needs of its consumers. However, it is a very fine line, where any attempt to use the moment to strengthen relationships could be fatal to the brand's reputation. It is essential that the brand focuses on providing genuine attention to consumers. Brand must stay true to its core values and purpose. The world, more than ever, demands brands' empathy. How might brands achieve empathy? Listening and talking closely to consumers (Garcia-Fuentes, 2020).

After a brand transgression, effective recovery must adequately restore relationship equity to customers and prevent negative online word of mouth from spreading. Personified brands must respond empathically and, if needed, explain the omission and offer an apology (Herhausen et al., 2019). Avoidance and nonresponsive behavior is the most inadequate response to restoring customer discomfort. Personified brands might establish a communication channel to provide sufficient and up-to-date information that helps consumers to cope with uncertainty. An active engagement brand behavior showing an empathic response is appropriate (Hill, Roggeveen, and Grewal, 2015). Empathic responses may enhance customer perception about brand courtesy, effort, and commitment. In affect-intensive environments such as the pandemic context, empathy represents a more effective approach (Homburg, Grozdanovic, and Klarmann, 2007). However, there is no recipe to follow. Theory and practice must come together in an attempt to offer solutions to a problem that has only just arrived. Brands need to adapt to the new normal consumption behaviors. Digital and social media marketing provide consumers with adaptive choice sets of products and services that ideally could lead to a lived experience of mindfulness, belonging, and happiness. As facilitators of intellectual capital, personified brands through social and digital media allow consumers to gain and circulate new knowledge that eventually will take them to greater well-being and mindful consumption. However, high emotional communications related to a social cause (e.g., pandemic) sometimes create suspicion about the brand's true commitment to the cause. In cases where brand social communication appeals to consumer guilt, low emotional intensity diminishes suspicion and stimulates consumer identification and positive brand perception (Singh et al., 2020). Brand social initiatives and communications may trigger attributional processes, such as attributions about the brand's motives, which may lead consumers to question its true intentions (Ellen, Webb, and Mohr, 2006). Consumer identification occurs when brand social initiatives imply values that are distinctive and relevant to consumers. The bigger the relevance, the greater the consumer's identification with the brand (Ahearne, Bhattacharya, and Gruen, 2005). When consumers believe that brand values resemble their self-image, they use the brand to demonstrate to themselves and society that their attitude, values, and behavior are approved by society. Brand identification enables customers to positively evaluate a brand, facilitating building a connection, loyalty, and trust. Brand trust comprises cognitive beliefs (e.g., expectation about the performance and quality of the brand) and emotional perceptions, which stem from emotional expectations about the brand's trustworthiness and benevolence. Brand trust has two dimensions: the first one is brand credibility as a product of cognitive beliefs. The second one is consumer protection related to the brand commitment to fulfill consumer needs (Hsu, 2019). Para Hsu (2019) brand identification positively affects brand trust, and brand trust positively impacts brand purchase. Customer interaction with brand communications leads first to involvement, being the subjective experience of consumers related to affective states, then to value perception and positive attitude toward the brand to finally influence purchase intention. Notably, consumers' identification with brands affects their support and ethical judgment for brands when committing a transgression. Consumers highly involved with identifying brands proactively defend the brand and shield it from other consumers' attacks (Dalman et al., 2019).

Personified brands need to act closely with consumers during crises like COVID-19 and show more empathy. However, it is not a simple task; to work effectively in online spaces, brand managers and consumers need to unlearn old behaviors to learn new ones that bring them together. Personified brands, as human-alike facilitators, face the challenge of guiding consumers through confinement defiances and engage them with new possibilities for change. Empathy allows brands to perceive a crisis from a different perspective, gain a more comprehensive panorama of its complexity, strengthen ties with consumers, and reinforce brand identification and loyalty. When brands, perceived as persons, show empathy, engage in meaningful communication with consumers looking for genuine and innovative solution-finding.

Managers must have full attention to how brands manage disruptive events that may affect their customer perceived and market value. Companies might also consider the strategic launch of supportive brand actions when being confronted with an unknown market situation to reinforce customer positive attitude towards the brand. International personified brand managers should be aware of consumers' higher likelihood of negative perception if their brand is seen as indifferent or uncaring.

Furthermore, advertising investment and main investments in selective partnerships help shield against negative customer perception online.

Further investigation might deepen in the social communication of personified brands as contributing to responsibility social marketing campaigns and ethical marketing where brand personification represents firm values encompassing functional, emotional, and social values.

KEYWORDS

- **brand empathy**
- **brand transgression**
- **COVID-19**
- **identification theory**
- **personified brands**
- **social theory**

REFERENCES

Aaker, J., Fournier, S., & Brasel. S. A., (2004). When good brands do bad. *Journal of Consumer Research, 31*(1), 1–16.

Ahearne, M., Bhattacharya, C. B., & Gruen. T., (2005). Antecedents and consequences of customer-company identification: Expanding the role of relationship marketing. *Journal of Applied Psychology, 90*(3), 574–585.

AMVO, (2020). *Estudio Sobre Venta Online en México 2020.* Retrieved from: https://www.amvo.org.mx/estudios/estudio-sobre-venta-online-en-mexico-2020/ (accessed on 18 February 2022).

Ang, T., Wei, S., & Anaza. N. A., (2018). Livestreaming vs. pre-recorded: How social viewing strategies impact consumers' viewing experiences and behavioral intentions. *European Journal of Marketing, 52*(9, 10), 2075–2104.

Aswani, R., Kar, A. K., Ilavarasan, P. V., & Dwivedi. Y. K., (2018). Search engine marketing is not all gold: Insights from Twitter and SEOClerks. *International Journal of Information Management, 38*(1), 107–116.

Belk, R. W., & Tumbat. G., (2005). The cult of Macintosh. *Consumption Markets & Culture, 8*(3), 205–217.

Bitner, M. J., Booms, B. H., & Tetreault. M. S., (1990). The service encounter: Diagnosing favorable and unfavorable incidents. *The Journal of Marketing, 54*, 71–84.

Copeland, M. T., (1923). *Relation of Consumers' Buying Habits to Marketing Methods* (Vol. 1, No. 3, pp. 282–289). Harvard Business Review.

Dalman, M. D., Buche, M. W., & Mi. J., (2019). The differential influence of identification on ethical judgment: The role of brand love. *Journal of Business Ethics, 158,* 875–891. doi: 10.1007/s10551-017-3774-1.

Day, G. S., (1969). A two-dimensional concept of brand loyalty. *Journal of Advertising Research, 9*, 29–36.

Elks, S., (2019). *Alexa, Are You Male or Female? "Sexist" Virtual Assistants Go Gender-Neutral.* Retrieved from: https://www.reuters.com/article/us-global-tech-women-idUSKBN1QS25B (accessed on 18 February 2022).

Ellen, Webb, & Mohr, (2006). Building corporate associations: Consumer attributions for corporate socially responsible programs. *Journal of the Academy of Marketing Science, 34*(2), 147–157.

Entrepreneur (2020). *Nike, Starbucks and Apple: The Latest Coronavirus Victims?* Retrieved from: https://www.entrepreneur.com/article/346007 (accessed on 18 February 2022).

Epley, N., & Waytz. A., (2010). Mind perception. In Fiske, S. T., Gilbert, D. T., & Lindzey. G., (eds.), *Handbook of Social Psychology* (pp. 498–541). New York: John Wiley & Sons.

Fournier, S., (1998). Consumers and their brands: Developing relationship theory in consumer research. *Journal of Consumer Research, 24*, 343–373.

Garcia-Fuentes, M., (2020). *¿Cuáles Son los Retos y Oportunidades de las Marcas Ante el COVID?* Entrepreneur. Retrieved from: https://www.entrepreneur.com/article/351124 (accessed on 18 February 2022).

Goldberg, L. R., (1993). The structure of phenotypic personality traits. *American Psychologist, 48*(1), 26–34.

Herhausen, D., Ludwig, S., Grewal, D., Wulf, J., & Schoegel. M., (2019). Detecting, preventing, and mitigating online firestorms in brand communities. *Journal of Marketing, 83*(3), 1–21. doi: 10.1177/0022242918822300.

Hill, K., Roggeveen, A., & Grewal. D., (2015). The impact of service recovery strategies on consumer responses: A conceptual model and meta-analysis. In: Diehl, K., & Yoon. C., (eds.), *Advances in Consumer Research* (Vol. 43, p. 789). Duluth, MN: Association for Consumer Research.

Homburg, C., Ehm, L., & Artz. M., (2015). Measuring and managing consumer sentiment in an online community environment. *Journal of Marketing Research, 52*(5), 629–641.

Hsu, L. C., (2019). Building brand-fan relationship in social commerce contexts: Mediators of online brand relationships. *Journal of Theoretical and Applied Electronic Commerce Research, 14*(2), 106–123. doi: 10.4067/S0718-18762019000200109.

Jacoby, J., (1971). Model of multi-brand loyalty. *Journal of Advertising Research, 11*(3), 25–31.

Kardes, F. R., (1988). Spontaneous inference processes in advertising. *Journal of Consumer Research, 15*(2), 225–233.

Karnes, K. C., (2020). *What is Emotional Branding and How to Use it Effectively*. Retrieved from: https://clevertap.com/blog/emotional-branding/ (accessed on 18 February 2022).

Khamitov, M., Grégoire, Y., & Suri. A., (2020). A systematic review of brand transgression, service failure recovery and product-harm crisis: Integration and guiding insights. *Journal of the Academy of Marketing Science, 48*, 519–542. doi: 10.1007/s11747-019-00679-1.

Kim, J., Allison, S. T., Eylon, D., Goethals, G. R., Markus, M. J., Hindle, S. M., & McGuire. H. A., (2008). Rooting for (and then abandoning) the underdog. *Journal of Applied Social Psychology. 38*, 2550–2573.

Krishen, A. S., & Berezan. O., (2019). Marketing and humanity: An introduction. In: Krishen, A. S., & Berezan. O., (eds.), *Marketing and Humanity: Discourses in the Real-World* (pp. 2–7). Newcastle upon Tyne: Cambridge Scholars Publishing.

Lin, J. S., & Sung. Y., (2014). Nothing can tear us apart: The effect of brand identity fusion in consumer-brand relationships. *Psychology & Marketing, 31*(1), 54–69.

Ma, L., (2020). When love becomes hate: How different consumer-brand relationships interact with crises to influence consumers' reactions. *Corporate Communications: An International Journal, 25*(3), 357–375. doi: 10.1108/CCIJ-08-2019-0103.
Mafi-Kreft, E., & Kreft. S., (2020). *Why Ford, Chanel, and Other Companies Pitch in During Crisis – Without the Government Ordering Them to.* Retrieved from: https://theconversation.com/why-ford-chanel-and-other-companies-pitch-in-during-a-crisis-without-the-government-ordering-them-to-135170 (accessed on 18 February 2022).
Magnusson, P., Krishnan, V., Westjohn, S. A., & Zdravkovic. S., (2014). The spillover effects of prototype brand transgressions on country image and related brands. *Journal of International Marketing, 22*, 21–38.
Nielsen, (2020). *COVID-19 Fuels a 50% Increase in Omnichannel Shopping Across the U.S.* Nielsen Insights, CPG FMCG & Retail. Retrieved from: https://nielseniq.com/global/en/insights/analysis/2020/covid-19-fuels-a-50-increase-in-omnichannel-shopping-across-the-u-s/ (accessed on 15 March 2022).
Obiegbu, C. J., & Larsen. G., (2020). Experiential brand loyalty: Towards an extended conceptualization of consumer allegiance to brands. *Marketing Theory, 20*(3), 251–271. doi: 10.1177/1470593119885167.
OECD (2020). *The Role of Online Platforms in Weathering the COVID-19 Shock.* Tackling Coronavirus (COVID-19). Contributing to a Global Effort. Retrieved from: https://www.oecd-ilibrary.org/economics/the-role-of-online-platforms-in-weathering-the-covid-19-shock_2a3b8434-en (accessed on 18 February 2022).
Oliver, R. L., (1997). *Satisfaction: A Behavioral Perspective on the Customer*. New York, NY: Routledge.
Paterson, M., & O´Malley, L., (2006). Brands, consumers, and relationships: A review. *Irish Marketing Review, 18*(1, 2), 10.
Pine, B. J., & Gilmore. J. H., (1999). *The Experience Economy: Work Is Theatre and Every Business a Stage.* Brighton, MA: Harvard Business School Press.
Regadera-Gonzalez, E., (2020). *Así Serán las Tiendas Después del Confinamiento.* The Conversation. Retrieved from: https://theconversation.com/la-moda-tras-la-covid-19-cambiaran-nuestros-habitos-de-consumo-137514 (accessed on 18 February 2022).
Sayin, E., & Gürhan-Canli. Z., (2015). Feeling attached to symbolic brands within the context of brand transgressions. In: *Brand Meaning Management* (pp. 233–256). Emerald.
Singh, J., Crisafulli, B., & Quamina. L. T., (2020). How intensity of journal of cause-related marketing. Guilt appeals influence consumers. The roles of company motives and consumer identification with the brand. *Advertising Research*, 148–162. doi: 10.2501/JAR-2018-049.
Stabler, S., & Fischer. M., (2020). When does corporate social irresponsibility become news? Evidence from more than 1,000 brand transgressions across five countries. *Journal of Marketing, 84*(3), 46, 47. https://doi.org/10.1177/0022242920911907.
Syrdal, H. A., & Briggs. E., (2018). Engagement with social media content: A qualitative exploration. *The Journal of Marketing Theory and Practice, 26*(1, 2), 4–22.

CHAPTER 2

Brand Engagement in Consumers' Lives During COVID-19 Times

CLAUDIA QUINTANILLA

EGADE Business School Monterrey, Mexico,
E-mail: Claudia.quintanilla@tec.mx

ABSTRACT

The COVID-19 pandemic has revealed that consumers cope with crises in different ways. This chapter presents a review of the brand engagement literature and the results of a qualitative study adopting a phenomenological approach based on 9 in-depth online interviews conducted with women, 19 to 50 years old, in a northeastern city of Mexico. The results show consumers' journey during the first two months of the pandemic, describing the new lifestyles generated by changes in consumption patterns. These lifestyles have changed the role and meaning of brands in their lives as well. Consumers have adopted new ways of shopping, new products, and new brands. Brand engagement strategies should be adapted to these new lifestyles as the brand consideration set, and the decision-making roles have been altered. Consumers have reorganized their priorities in life: health first, with no more concerns about fashion, learning that "less is more."

2.1 INTRODUCTION

In mid-March 2020, a typical day for most people included commuting to work, shopping for groceries, picking up their children from school,

Marketing by Contingency in the Time of COVID-19: Overcoming Business Crises and Meeting Marketing Challenges. Alicia De La Peña, PhD & Juan Bernardo Amezcua Nunez, PhD (Eds.)

and going out to buy a treat at the corner store—all daily activities that we never imagined we would ever miss. Suddenly people's lives were altered entirely because of the global COVID-19 pandemic lockdown, which required social distancing and strong sanitary regulations. Government regulations to control the pandemic, which most expected to last a few days or maybe a few weeks at most, have remained in place for more than 45 weeks. Such regulations included self-isolating at home as much as possible, leaving only to shop for essential supplies, and practicing social distancing by maintaining a distance of at least 1.5–2 meters from others. This chapter aims to contribute to consumer behavior knowledge by documenting the changes in consumers' lives and the role that brands have played during this lockdown.

According to J. Walker Smith, chief knowledge officer (CKO) of Kantar (a leading global consultancy in data, insights, and consulting), in the "new normal"—referring to the new lifestyle standard—established and strong brands will be those that remain in the market (AMA Conference, 2020). A comparative study of the brand value ranking indexes, which track trends over time using information from the S&P500, BrandZ, and MSCI, provided evidence of this statement. Charles Trevail, CEO of Interbrand, also pointed out that, in the global report of the best brands for 2020, despite the turbulence generated by COVID-19, the added value of the 100 best brands in these global rankings grew by 9% in 2020 (Interbrand, 2020). Brands play an important role in consumers' lives both during and without a pandemic, as over time, they build brand equity for the products and services they are endorsing, becoming readily recognized by their loyal users. Consumers have a relationship with brands, as Keller (2020) explained that brands not only make consumers' lives easier or more rewarding but also allow consumers to signal to others or themselves who they are or who they want to be. Brand equity has been defined by a variety of approaches, classified by Feldwick (1996) as: (i) the total value of a brand as a separable asset when it is sold or included on a balance sheet; (ii) a measure of the strength of consumers' attachment to a brand; and (iii) a description of associations and beliefs the consumer has about the brand. The brand value built through the marketing efforts oriented to develop positive associations among consumers and brands are not only tangibles but are also considered intangibles, as Dibb et al. (1997) indicated in their modification to the definition of brand by Bennett (1988, p. 18): "A brand is a name, term, design, symbol or *any other feature* that identifies one seller's good or service as distinct from those of other sellers." However, over time we see new brands being born while others die because they do not have the necessary elements to develop this brand equity for their customers and are not able to achieve a strong brand engagement.

What are brands expected to bring to the market? Initially, brands functioned as a quality signal of products, meaning a good brand was a guarantee of quality, durability, and expertise. Over time, brands shifted to a focus on people, developing the concept of brand personality; as a result, they became an extension of their users to communicate that the consumers who use those brands identify with the brands' personality (Aaker, 1997). However, as a society, we have evolved and aspire to be better, so we now demand that brands contribute in some way to building a better society. Through this evolution, the marketing strategies developed first focused on the quality of products or services, then on the audience or consumers and users, and finally on the community. This is how brands satisfy their consumers: They build loyalty and develop love, commitment, and brand engagement.

This chapter is based on research that explores consumption patterns and experiences with brands during the first 60 days of the COVID-19 lockdown. The first section presents a literature review of engagement, including customer engagement and customer brand engagement (CBE), to understand the functions and relevance of brands in consumers' lives; also, a section about consumption and economic and health crisis is developed. The subsequent section explains the methodology used to answer the research question, presents participants' profiles, and explains the data analysis process. The results section details the findings classified into four themes: (a) new shopping habits and decision-making roles, (b) coping strategies for a new way of life, (c) role of brands during lockdown, and (d) brands after isolation. Finally, the conclusion summarizes the insights after the interpretative analysis of the data.

2.2 CUSTOMER BRAND ENGAGEMENT (CBE)

Engagement is generally defined as "a formal agreement to get married" (Cambridge Dictionary); this agreement represents a strong commitment among the actors involved. Some contexts in which engagement has been studied are education as student engagement (Fredricks et al., 2004;

London et al., 2007), organizational as employee engagement (Schaufeli et al., 2002), and sociology as civic engagement (Jennings and Zeitner, 2003).

The engagement concept has also been explored from the marketing perspective. The customer engagement framework proposed by Bowden (2009) states that, historically, customer satisfaction was the standard measurement for assessing customers' responses to products and services based on the belief that high levels of satisfaction may increase loyalty and purchase intention. Yet satisfaction theory based on the confirmation disconfirmation of expectations approach has been highly criticized because it is based on a cognitive post-consumption approach (Bartikowski and Llosa, 2004; Parasuraman et al., 1988; Wirtz et al., 2000), and the concept omitted emotional and behavioral dimensions. Therefore, Bowden (2009) proposed redirecting satisfaction research toward a construct that includes commitment, involvement, and trust in the creation of engaged and loyal customers while considering concepts studied by McEwen (2004), who argued that engagement should include confidence, integrity, pride, and passion in a brand. Based on this framework, a more holistic construct is proposed: CBE, which is defined as "the level of a customer's cognitive, emotional, and behavioral investment in specific brand interactions" (Hollebeek, 2011, p. 555). Hollebeek (2011) further identified three themes that connect directly to each of the dimensions included in the definition: immersion, passion, and activation. A strong antecedent of CBE is the consumer involvement construct (Dwivedi, 2015; Leckie et al., 2016), which indicates that a brand is relevant for a consumer (attitudinal dimension), who decides to spend time, money, and effort (behavioral dimension) to develop a relationship with the brand over time (Harrigan et al., 2017). Reviewing marketing literature on CBE, relationships with brands are based on cognitive, emotional, and behavioral dimensions. Although recent research about engagement during the COVID-19 pandemic has proposed paradigmatic lenses to understand and probably redefine the concept, based on human interactions, three lenses are presented: economic rationalism, institutionalism, and existential-humanism (Karpen and Conduit, 2020). The economic rationalism lens defines the consumer's goal to maximize return in the investment and usage from the product or service acquired. In contrast, institutionalism centers on social elements and structures that enable connections among consumers (Sprott et al., 2009). Finally, existential-humanism refers to the self-reflection of questioning the state of being and living (Berry, 2020). Customers look for a reflection of the personal values they practice in their brands. As Arora et al. (2020) stated, customers in uncertain times seek to engage with businesses in different ways than the way they used to before, and perhaps the call for engagement research proposed by Conduit et al. (2019) to study the deeper sense of being and meaning as the spiritual dimension is relevant more than ever.

2.3 CONSUMPTION AND CRISIS

In times of economic or health crisis, consumers are affected by situations that cause fear and uncertainty in the environment, and this affects how consumption decisions are made and purchasing patterns change. Economic crises generally represent consumers' concern over work and household sources of income loss. COVID-19 pandemic has not been an exception and has generated a global economic crisis with its own characteristics compared with the Great Financial Crises during 2007–2009 (Borio, 2020). According to Nielsen, during the 2009 economic crisis in Mexico, consumers responded in different ways: (a) the consumer needed to decide what to spend on; (b) the consumer decided to switch to brands of lower price and generate savings; switching to private labels, (c) the households with greater number of members and with lower income had been forced to rationalize more, this is also reflected in the migration towards smaller packages which leads to towards a greater frequency purchase. Despite the above, people with low purchasing power are more sensitive to sales, so they were bombarded with promotions of cheaper or smaller packaging quantities (AC Nielsen Mexico, 2009). In contrast, during the COVID-19 economic crisis, systems in digitally advanced countries, such as China, were already prepared for contactless delivery, augmented reality-enabled shopping, mobile, and tap-to-pay payments. Even though in Mexico, companies were not yet at the digitalization level of China, Mexican Online Sales Association (AMVO) reported in June 2020, only 6 out of 10 SMEs were selling online, which represented an increase of 94.6% compared to 2019; despite the negative impact of the pandemic, digital sales grew by more than 10%.

Another common reaction to a health crisis is people making panic purchases. This caused a shortage of many products in supermarkets, and thus consumers were led to try products they had not considered before. There is evidence of product categories affected positively or negatively in previous health crises, i.e., H1N1 or Influenza in 2009 and SARS in 2003. Differences in buying behavior in the different pandemics that have been generated in China with SARS, Mexico with Influenza and Coronavirus evidence that while in China there was no panic shopping behaviors during SARS 2003 and COVID-19; in Spain and Mexico during COVID-19, consumers make panic shopping looking for cleaners, hand soap, detergents. (PMfarm, 2020). Also, according to Castaño, Quintanilla, and Perez (2010), Mexican consumers during Influenza in 2009 when retail was not yet as digitalize as it is now, consumption patterns change to cheaper brands and wholesale purchases from the middle socioeconomic segment, while lower socioeconomic consumers look for sales and private brands.

Therefore, this study aims to explore the role that brands play in a health crisis situation with isolation such as that generated by the pandemic. Given the vulnerability shown globally during a pandemic like COVID-19, it cannot be ruled out that, in the future, there will be similar conditions again, which is why it is relevant to learn about consumers' reactions and especially the role that brands play in their lives to comply with the brand's community function mentioned in the introduction of this chapter. Brands looking to prevail in the market must develop an appropriate engagement level with their customers.

2.4 METHODOLOGY

A qualitative methodology was used to explore changes in consumer behavior, perception, and meaning of brands in a time of confinement, such as that generated by COVID-19. A qualitative methodology was used precisely to delve into the participants' motivations, meanings, and experiences through the details of their narratives on the phenomenological approach, nine in-depth interviews were conducted with women of medium-high socioeconomic status between 18 and 50 years old in a city in northwestern Mexico during the last two weeks of May 2020—a relevant period for the study as they had completed at least two months of isolation. Following theoretical saturation criteria recommended for qualitative research, the final sample was nine in-depth interviews (Boddy, 2016). All interviewees reported economic stability at their homes up to the fieldwork data. Table 2.1 presents participants' profile descriptions.

TABLE 2.1 Participants Profile

Participant	Name	Age	Marital Status	Living Situation
1	Valeria	19	Single	Parents 1 sister
2	Vicky	23	Single	Parents 2 sister
3	Dany	30	Married	Husband 1 kid (4 years old)
4	Miriam	32	Married	Husband 2 kids (4 and 2 years old) 1 Domestic employee
5	Denise	33	Single	Parents 1 brother
6	Alejandra	38	Married	Husband 2 pets: cat and dog
7	Norma	45	Married	Husband 2 kids (10 and 4 years old)
8	Rebecca	46	Married	Husband
9	Diana	50	Single	2 kids (9 and 3 years old) Lives alone 1 pet: dog

Source: Own elaboration.

The phenomenological approach was used as the main objective is to study the life experience during the phenomenon from the interviewee's perspective (Thompson et al., 1990). During this time of the pandemic, several quantitative studies have been carried out to describe and measure the changes in consumption patterns; however, being able to explore the motivations, meanings, and consumers' internalization of these experiences is what justifies the qualitative methodology used in this study.

Through a purposeful sampling method, participants were recruited based on who could provide relevant, diverse, and pertinent information to the research topic (Tong et al., 2007), including a filter question on the number of days they had been in isolation, where the response should be at least 60 days. Age dispersion was sought, and only women were selected because there is evidence that shows that women tend to develop a stronger engagement with brands (Sherrod, 1989). Likewise, in Mexico, women have a greater responsibility for household purchases; during the pandemic, this responsibility has been emphasized due to restrictive measures of access to places of purchase.

In-depth interviews allow for the study of human complexity through immersion in the reflections, internal thoughts, and feelings of another person through a format of immediate interaction between the interviewee and the interviewer (Salmons, 2014). The interviews were conducted synchronously online through the Zoom platform, which allowed them to be recorded with the participants' consent. The interviews were also carried out from an emic perspective, where the researcher tries to capture—as an "insider" of the group under study—the events of the real world (Yin, 2016, p. 11) and looks at things through the eyes of the members of the culture being studied (Willis et al., 2007, p. 100). A semi-structured interview guide was designed to be applied in the same order to all participants and included open questions organized into four sections: isolation and consumption; isolation and brands; visualization after quarantine; and questions that contextualized the confinement and accompanying levels as well as demographic variables. A semi-structured interview allows the interviewer to use the same wording in each interview and reduces possible biases that may be generated (Salmons, 2014). To establish a good rapport, the researcher initiated projective questions like "a day in your life in a time before isolation," continuing with what they missed most about their previous way of life. The researcher also asked if for some reason, the participants had broken the rule of staying at home, for what reason, and if it was worth it. Throughout this process, the researcher showed attention and great interest in participants' answers to generate an atmosphere of trust and empathy between the interviewer and the interviewee (Wolgemuth et al., 2015). Each interview lasted approximately 45 minutes, and 97 pages of transcripts were generated. Each transcript was analyzed using a data manipulation process to identify emerging categories, organize themes, generate codes, categorize, and extract meanings (Adu, 2019).

2.5 RESULTS

Documenting participants' experiences during isolation was a relevant process of this research. As a result of the projective technique, "a day of your life before the confinement," when compared with the current way of life, participants identified feelings of loneliness, nostalgia, and even anxiety and depression generated by the uncertainty about how long the confinement will last. They shared the changes they have experienced in their consumption patterns and the role that brands have had during these months in this new way of life.

2.5.1 NEW SHOPPING HABITS AND DECISION-MAKING ROLES

Participants mentioned that their shopping habits had changed drastically. They had become first-time online shoppers in categories that they had not been used to buying online before. Although all participants reported not losing their income source as of the date of this study, they recognized a more reasoned and conscious way of shopping characterized their decisions as they searched for "essentials" and engaged in less impulsive behavior.

"We now buy groceries at Walmart online, and we can compare prices on a lot of websites, not only there" (Miriam, 32 years old).

"I am only buying the essentials. I no longer go for cravings… crazy or compulsive shopping… not anymore" (Diana, 50 years old).

"Now I need to plan even the food cravings to buy at the supermarket, so they can be available whenever we need it, and we don't buy them in the convenience store as we used to" (Norma, 45 years old).

"When I buy the groceries online, I also buy food for our cravings…. I ask my husband what he would like to have to meet his cravings; before COVID-19… he used to go to OXXO (convenience store) any time he needed" (Alejandra, 38 years old).

For those participants for whom buying online is not a feasible alternative because they have not gone digital yet, the number of visits to the supermarket has decreased. For families who have the purchasing power to buy food to cover more extended periods of time, they now acquired larger packages compared to before the pandemic.

"We now have this proactive attitude that COVID-19 left us… we go shopping once per month… it is cheaper, and we are no longer exposed (to contagious people) every weekend" (Dany, 30 years old).

For online shoppers who purchased products in new categories, consumers felt empowered to make more informed buying decisions, learning about the relevant attributes to be considered in their selection.

"We want to buy a trampoline for the entire family to exercise. Before the pandemic, I went to the store, and only I evaluated the options there versus my budget; but buying online allows me to learn about the relevant attributes I should consider in that product evaluation and also, I have the chance to compare different stores" (Miriam, 32 years old).

Family members became more involved in in-house chores during the lockdown, which was another reason to alter the decision-making process. Usually, the female household decided and selected the brands to consume, but during the COVID-19 pandemic, new brand relationships were developed among family members not committed to those activities and did not even consider the existence of those brands before the lockdown.

"Before COVID-19, I bought clothes for the kids; now my husband asked me if the kids really need more clothes, if I'm sure they really don't have anything to wear" (Miriam, 32 years old).

"Before the pandemic, my husband was never involved with the bathroom cleaning brand… now he has a relationship with it … [he said] 'don't change that cleaner anymore!'" (Dany, 30 years old).

"Now my husband is aware if the number of tomatoes is decreasing… he does not only care about the beer and snacks inventory like before this pandemic situation" (Alejandra, 38 years old).

2.5.2 COPING STRATEGIES FOR A NEW WAY OF LIFE

Participants mentioned that the changes in their daily routine generated confusion, stress, and fatigue. To cope with the feelings of loneliness, depression, and anxiety caused by this change in life, participants developed different coping strategies—namely, social, occupational, and therapeutic strategies. Some examples of these strategies include establishing a routine and following a schedule, socializing with friends or family members in a face-to-face or digital format, practicing new activities (e.g., exercising, baking, walking the pet, painting), seeking professional help from a therapist, and developing a short-term mindset (see Table 2.2).

2.6 ROLE OF BRANDS DURING LOCKDOWN

2.6.1 MISSED BRANDS

Participants missed brands associated with important routines or rituals in their lives. The brands play an important role in special moments, including moments for themselves and time shared with their loved ones. They miss their brands because of the strong engagement they have with them. This isolation provoked nostalgia, and they recognized that they had not broken their brand relationships; only those brands are in "standby" mode. Missed brands were also related to categories of products and services that represent routines and/or rituals in consumers' lives. One highly missed brand mentioned by 4 out of 10 interviewees was the iconic brand Starbucks. Some reflections mentioned by participants about missed brands follow:

TABLE 2.2 Social, Occupational, and Therapeutic Coping Strategies

Occupational	"I stick to my routine, I walk up at the same time as if I were going to the office; I get dressed and put on makeup…the only thing different is the shoes—I wear tennis shoes and I am comfortable" (Alejandra, 38 years old).
	"I walk our dog every day and do my exercise routine" (Alejandra, 38 years old).
	"Catching up with the oil painting hobby I used to practice, downloading TikTok app…these have been useful for not wasting the time, but using it instead of just watching TV series" (Valeria, 19 years old).
	"I've start cooking for my family, a sweet potato and banana pie frequently…I also started doing clay figures" (Vicky 23 years old).
Social	"We continue to have family gathering…and that helps us to socialize. Grilling on weekends" (Denisse, 33 years old).
	"I have had Zoom meeting with friends, with the former neighbors. At least…I know how they are…but nothing compares to the 'pat on the back'…referring to physical contact" (Diana, 50 years old).
	"With the family members that live outside the city…we have Zoom meeting on Sundays after lunch…and we catch up about COVID-19 status in each city" (Valeria, 19 years old).
	"Talking to my family calms me down. As a family we have 'movie night'…and we went out to the patio…all together to watch a movie and we relaxed" (Vicky, 23 years old).
Therapeutic	"Praying the rosary…it gives me peace from all the global uncertainty…and I am also in therapy with a psychologist" (Miriam, 32 years old).
	"If I stop thinking in the long term, my days go by very quickly" (Dany, 30 years old).

Source: Own elaboration.

"I miss Starbucks because it was part of my routine at work. It gave me a relaxed feeling" (Alejandra, 38 years old).

"Starbucks—I miss the experience of going there… I miss the taste, of ordering the same every time I go and being able to sit down and talk for a while with my boyfriend or friends" (Denise, 33 years old).

"Starbucks brand is what I miss… I miss that time for myself… when I left my kids at kindergarten and went to pick up my daily treat" (Miriam, 32 years old).

"Starbucks represents my daily treat. It was that part of my daily routine, something I enjoy in my loneliness… the few moments a day that I have the chance to be by myself" (Dany, 30 years old).

"I miss Chilis because I used to meet different groups of friends there… it is the companions that make me miss the brand" (Diana, 50 years old).

"Brands that are my identity are with me during the quarantine" (Denise, 33 years old).

"I want to do my normal life… I want to go to the San Agustin Gym" (Vicky, 23 years old).

"Stradivarius and Lefties…I feel anxious that once the quarantine is lifted, I want to be able to go see what's new… we've already lost many months of fashion" (Valeria, 19 years old).

"I used to say I'm going to go to H&M just to look around. Now it is not on my mind anymore" (Dany, 30 years old).

For some missed brands, the resulting gap in consumers' lives is now filled with activities rather than with other brands. These activities allow consumers to maintain a connection with themselves or their beloved ones.

"I can't say they are not loved brands by me anymore… they are in a standby situation, because I am not using them right now. I generated a distance with them, but they are not out of my life" (Dany, 30 years old).

"I am spending the time I used to go for my Starbucks to cook with my husband… a gourmet breakfast. We talk about what will happen during the day. These activities allow us to be more relaxed and united" (Alejandra, 38 years old).

"I have tried to replace Starbucks because now in confinement, I do my routine and say mid-morning I am going to have a tea with honey… I love it!!! I try to drink it when I am not with the kids" (Norma, 45 years old).

2.6.2 NEW BRANDS

Participants mentioned that they have developed new relationships with new brands for different reasons: (a) the brand is out of stock because of the lockdown and the supplier's logistic problems, and (b) they want to support local retailers because of the economic situation. Their awareness of others' situations makes them decide more in terms of community and not necessarily think of their self-interests.

"We started buying a new ham brand since the one we like is imported and out of stock" (Alejandra, 33 years old).

"I've started using a new shampoo, because I can't find my brand on the website where I now buy online" (Rebecca, 46 years old).

"We started shopping at a new little grocery store near home, since they set up home delivery service and it is easier, nearby, and we can help a family firm in these hard economic times" (Norma, 43 years old).

"At home we started buying prepared meals from our neighbors to help them, since they lost their job" (Diana, 50 years old).

2.7 BRANDS AFTER ISOLATION

After isolation, participants envision themselves spending time with their families and friends, with the hope of not being afraid of getting infected anymore while enjoying the sense of freedom they remember from before isolation. They are nostalgic about activities such as going to friends' houses, sharing time at restaurants, traveling, and sharing experiences.

Participants recognized that they had developed a self-reflection about the "global impact," meaning, realizing that geographic distance is not as big as they thought before the pandemic. In addition, they have developed a new understanding of what is valuable for them and what they should pay attention to in terms of consumption. They have reorganized their priorities in life: Health comes first, and they are no longer concerned about fashion. They have learned the lesson that "less is more."

"This time has made me reflect, how easy it is to be well, to have a good time… or to feel pretty… I don't need a lot of stuff on me anymore" (Dany, 30 years old).

"Examples of people supporting vulnerable people have taught me that there are still good people in the world" (Alejandra, 38 years old).

"The routine will change after isolation. I will become more aware that we must take advantage of certain moments... take advantage of outings... to value many simple things that you take for granted" (Denise, 38 years old).

"I have realized that I do not need so many clothes and shoes to live" (Diana, 50 years old).

"I am more aware of the impact that each of us can make on others. This situation started so far... in China and look how it has affected our lives" (Vicky, 23 years old).

2.8 CONCLUSION

This chapter aims to contribute to the understanding of how consumers' lifestyles and their consumption patterns have changed during the lockdown due to the pandemic of COVID-19. Participants in this study had completed two months in isolation, following government recommendations. The study sought to document the strategies developed to cope with this big change in their lifestyles, and the role brands have had. The pandemic had pushed them to try shopping in new ways; more than half of them stated that they had recently ordered groceries online for the first time and purchased other categories they have never bought online before. Brands have played an essential role during isolation, and, of course, the brand consideration sets have been altered. Newly discovered brands might become part of consumers' lives due to the good experiences provided, even considering that consumers were initially "forced" to try them because their usual brands were unavailable. Some have found new brands or product substitutions for specific brands and even appear to be adopting "new lifestyles."

The decision-making roles have also been altered as consumers have spent more time at home. Some members' participation in chores that they did not usually do has increased through their involvement in shopping decisions. With more time at home, consumers have been dedicating their time in isolation to activities that help them feel better, such as practicing new and old hobbies (photography, reading, painting, cooking, knitting, etc.). Some of these activities are practiced to feel better. In contrast, other activities are undertaken to replace the brands and are related to the meaning the brands had in their lives before isolation. These activities allow them to have their own moments to themselves, pampering themselves, providing self-care, or sharing time with their family.

The findings revealed that consumers missed the brands and the routines or rituals that evoke the consumption of these brands, so the question is: How can brands generate meaningful personal experiences in the new context? Brands must relearn what they knew about their consumers' behavior in order to continue engaging with them through the development of experiences that allow consumers to interact with others by taking advantage of technology and rethinking the value proposition offered before the pandemic. For example, parents of children in kindergarten paid not only for an educational and value system for their children but also for the time that their children interacted with others in a space outside the home, socializing with the teacher and their peers. Distance education for kindergarteners in this new reality makes parents wonder what value proposition they are really acquiring because, in the new reality, they are doing all the activities the teacher used to do themselves. As in this example, many other categories of products and services must study and redesign their value propositions for this "new normal" context.

Consumers want to disconnect from this new reality and not think about the long term anymore; in this sense, brands must rethink the social function they fulfill. Some brands' role is to communicate self-expression and self-presentation (Wilcox et al., 2009). According to the insights from this study, some brands are kept in the closet or inside the house in a "standby" mode. How can brands continue to be relevant for their consumers? Brands in "standby" mode must develop strategies to survive in the new reality. Consumers recognize and value that brands—during the pandemic and always—help their lives, worry about their health safety and also that of their employees, adapt the point of sale, and develop e-commerce options and apps that allow them to continue consuming. Brands need to identify new social networks, influencers, and communication means as well as create a relevant message to reconnect with their customers and maintain the engagement they have already built.

KEYWORDS

- **brand engagement**
- **brands**
- **consumer behavior**
- **consumption patterns**
- **COVID-19**
- **engagement**

REFERENCES

Aaker, J. L., (1997). Dimensions of brand personality. *Journal of Marketing Research, 34*, 347–356.

AC Nielsen, (2009). *Global Study of Consumer Confidence in the Face of the Global Economic Crisis* [Press Release]. https://www.nielsen.com/latam/es/insights/ (accessed on 18 February 2022).

Adu, P., (2019). *A Step-by-Step Guide to Qualitative Data Coding*. Routledge.

Arora, N., Charm, T., Grimmelt, A., Ortega, M., Robinson, K., Sexauer, C., & Yamakawa. N., (2020). *A Global View of How Consumer Behavior is Changing Amid COVID-19*. McKinsey and Company.

Bartikowski, B., & Llosa. S., (2004). Customer satisfaction measurement: Comparing four methods of attribute categorizations. *The Service Industries Journal, 24*(4), 67–82.

Bennett, P. D., (1988). *Dictionary of Marketing Terms*. The American Marketing Association.

Berry, W., (2020). *The Existential Crisis You Are (or Should Be) Having*. Psychology Today. https://www.psychologytoday.com/us/blog/the-second-noble-truth/202004/the-existential-crisis-you-are-or-should-be-having (accessed on 18 February 2022).

Bowden, J. L. H., (2009). The process of customer engagement: A conceptual framework. *Journal of Marketing Theory and Practice, 17*(1), 63–74.

Cambridge University Press, (n.d.). *Upcycling*. In Cambridge dictionary. Retrieved from: https://dictionary.cambridge.org/us/dictionary/english/engagement (accessed on 18 February 2022).

Castaño, R., Quintanilla, C., & Perez. M., (2010). Attitudes and behavior of Mexican female consumers: Their reactions to crisis situations. *XIV International Congress of the Academy of Administrative Sciences AC (ACACIA)*.

Conduit, J., Karpen, I. O., & Tierney. K. D., (2019). "Volunteer engagement: Conceptual extensions and value-in-context outcomes." *Journal of Service Theory and Practice, 29*(4), 462–487.

Dibb, S., Simkin, L., Pride, W. M., & Ferrell. O. C., (1997). *Marketing: Concepts and Strategies* (3rd European edn.). Houghton Mifflin.

Dwivedi, A., (2015). A higher-order model of consumer brand engagement and its impact on loyalty intentions. *Journal of Retailing and Consumer Services, 24*, 100–109.

Feldwick, P., (1996). Do we really need "brand equity?". *The Journal of Brand Management, 4*(1), 9–28.

Fredricks, J. A., Blumenfeld, P. C., & Paris. A. H., (2004). School engagement: Potential of the concept, state of the evidence. *Review of Educational Research, 74*, 59–109.

Harrigan, P., Evers, U., Miles, M., & Daly. T., (2017). Customer engagement with tourism social media brands. *Tourism Management, 59*, 597–609.

Hollebeek, L., (2011). Exploring customer brand engagement: Definition and themes. *Journal of Strategic Marketing, 19*(7), 555–573.

Interbrand (2020). Best global brands 2020: A New Decade of Possibility. Retrieved [7-01-2020] Available at: http://interbrand.com/wp-content/uploads/2020/BGB_2020_Report.pdf (accessed on 14 March 2022).

Jennings, K. M., & Zeitner. V., (2003). Internet use and civic engagement: A longitudinal analysis. *Public Opinion Quarterly, 67*, 311–334.

Karpen, I. O., & Conduit. J., (2020). Engaging in times of COVID-19 and beyond: Theorizing customer engagement through different paradigmatic lenses. *Journal of Service Management.*

Keller, K. L., (2020). Consumer research insights on brands and branding: A JCR curation. *Journal of Consumer Research, 46*(5), 995–1001.

Leckie, C., Nyadzayo, M. W., & Johnson. L. W., (2016). Antecedents of consumer brand engagement and brand loyalty. *Journal of Marketing Management, 32*(5, 6), 558–578.

London, B., Downey, G., & Mace. S., (2007). Psychological theories of educational engagement: A multi-method approach to studying individual engagement and institutional change. *Vand. L. Rev., 60*, 455.

McEwen, W. J., (2004). Getting emotional about brands. *Gallup Management Journal, 9*, 1–4.

Parasuraman, A., Zeithaml, V. A., & Berry. L. L., (1988). Servqual: A multiple-item scale for measuring consumer perc. *Journal of Retailing, 64*(1), 12.

PMfarma, (2020). *Impacto en el Consumo Masivo Durante Pandemias. ¿Cómo las Marcas Pueden Mitigar los Efectos Negativos del Coronavirus*? Retrieved from: https://www.pmfarma.com.mx/noticias/16594-impacto-en-el-consumo-masivo-durante-pandemias.-como-las-marcas-pueden-mitigar-los-efectos-negativos-del-coronavirus.html (accessed on 18 February 2022).

Salmons, J., (2014). *Qualitative Online Interviews: Strategies, Design, and Skills*. Sage Publications.

Schaufeli, W. B., Martínez, I. M., Pinto, A. M., Salanova, M., & Bakker. A. B., (2002). Burnout and engagement in university students: A cross-national study. *Journal of Cross-Cultural Psychology, 33*, 464–481.

Sherrod, D., (1989). The influence of gender on same-sex friendships. *Close Relationships, 10*, 164–186.

Sprott, D., Czellar, S., & Spangenberg. E., (2009). The importance of a general measure of brand engagement on market behavior: Development and validation of a scale. *Journal of Marketing Research, 46*(1), 92–104.

Thompson, C. J., Locander, W. B., & Pollio. H. R., (1990). The lived meaning of free choice: An existential-phenomenological description of everyday consumer experiences of contemporary married women. *Journal of Consumer Research, 17*(3), 346–361.

Tong, A., Sainsbury, P., & Craig. J., (2007). Consolidated criteria for reporting qualitative research (COREQ): A 32-item checklist for interviews and focus groups. *International Journal for Quality in Health Care, 19*(6), 349–357.

Wilcox, K., Kim, H. M., & Sen. S., (2009). Why do consumers buy counterfeit luxury brands?. *Journal of Marketing Research, 46*(2), 247–259.

Willis, J. W., Jost, M., & Nilakanta. R., (2007). *Foundations of Qualitative Research: Interpretive and Critical Approaches*. Sage.

Wirtz, J., Mattila, A. S., & Tan. R. L., (2000). The moderating role of target-arousal on the impact of affect on satisfaction—an examination in the context of service experiences. *Journal of Retailing, 76*(3), 347–365.

Wolgemuth, J. R., Erdil-Moody, Z., Opsal, T., Cross, J. E., Kaanta, T., Dickmann, E. M., & Colomer. S., (2015). Participants' experiences of the qualitative interview: Considering the importance of research paradigms. *Qualitative Research, 15*(3), 351–372.

Yin, R. K., (2016). *Qualitative Research from Start to Finish* (2nd edn.). The Guilford Press.

CHAPTER 3

Marketing in a Box: Resilient Strategies Amid the Crisis

ALICIA DE LA PEÑA

School of Marketing, Universidad Autónoma de Coahuila, Mexico, E-mail: aliciadelapenadeleon@uadec.edu.mx

ABSTRACT

From mystery boxes to promote touristic destinations to snow day kits, both goods and services firms have been using containers and boxes as a marketing tool to deliver their offers right into the hands of their consumers during the COVID-19 lockdown. After governments around the World announced massive lockdown measures to prevent the spread of the COVID-19 virus in March 2020, marketers had to use their creativity to keep business afloat. It is well known that the COVID-19 pandemic affected traditional business models. In the early months of the pandemic, many companies remained closed, losing sales and the opportunity to help individuals face the situation with their favorite brands at hand. Those businesses that already had an established e-commerce model before the pandemic began were definitely more resilient and able to serve their customers by selling online. But, what happened with small enterprises and service firms that were not using e-commerce strategies before COVID-19? Did they close their doors for good, or did they creatively design new forms to reach their old customers and attract new ones? This chapter explores some of the strategies adopted by different firms to remain relevant during the COVID-19 pandemic, demonstrating that those

Marketing by Contingency in the Time of COVID-19: Overcoming Business Crises and Meeting Marketing Challenges. Alicia De La Peña, PhD & Juan Bernardo Amezcua Nunez, PhD (Eds.)

enterprises that want to survive a crisis need to think out of the box and present new and creative offers.

3.1 INTRODUCTION

It is well known that the COVID-19 pandemic affected traditional business models. For most businesses, in-person transactions suddenly became a rare event. During the first weeks of the lockdown and before extreme measures were taken by authorities, consumers ran to the stores and began to stockpile what they thought would be needed to survive for a couple of weeks: frozen and canned food became a must in every shopping list, as well as toilette paper, antibacterial gel, face masks, and disinfecting and cleaning supplies (Sheth, 2020).

But the COVID-19 lockdown that seemed to be a short-term event soon became the new normal, and consumers had to modify their shopping patterns and lifestyles, not for a couple of weeks, as they initially thought, but for several months. Some, very afraid to get infected by the virus, decided to turn to digital technologies and use their smartphones and store apps to buy groceries, office supplies, ready to eat meals, and even coffee from their favorite coffee parlor; opening the door of their homes just to pick up the deliveries that were handed to them by face covered deliverymen. Others kept some of their old shopping habits, still went to the stores, but followed regulations such as keeping the social distance, wearing masks, and buying limited quantities of some items (Barnes et al., 2020; Pantano et al., 2020).

During the lockdown, anxiety and stress became the normal mental state for many. Still, shopping either online or at brick and mortar stores helped individuals cope with the negative emotions, and their favorite brands became the analgesic that enabled them to soothe their distress and find some sort of company (Laato et al., 2020; Reimann et al., 2017). Moved by misinformation and seeking to release the panic they felt, many began to stockpile food, toilet paper, and cleaning items (Barnes, Diaz, and Arnaboldi, 2021; Kirk and Rifkin, 2020). It was buying supplies to equip their home fortress that helped families build their own home offices and home schools and kill the free time and fight the boredom resulting from being day and night locked in the same place (Marketing Charts, 2020). While many decided to turn to Instagram, Tik Tok, or YouTube to get ideas on how to use the artifacts and supplies they already owned to cook, learn, work or entertain at home, others went online surfing seeking for new products and ideas to solve not only their basic needs but also looking for replacements to fill the void resulting from the isolation and the lack of social encounters (Gammon, 2020; Kantar, 2020).

Those changes in consumption patterns represented a challenge for most companies. First, hypermarkets, grocery stores, pharmacies, and convenience stores had to overstock disinfecting and cleaning products to cover the growing demand of anxious and scared consumers. Then, retailment and merchandising strategies were modified to comply with new regulations and provide both employees and customers with a safe environment in the store (Pantano et al., 2020; PWC, 2020).

In Mexico, as well as in many other countries, businesses were divided into essential and non-essential. The essential companies included, among others, supermarkets, pharmacies, banks, gas stations, hardware stores, laundries, funeral houses, shipping outlets, auto repair, and restaurants with takeout services (CDMX, 2020). At the same time, all businesses related to industries such as entertainment, recreation, and hospitality, including gyms, spas, concert venues, movie theaters and restaurants and bars with dine-in or drink-in services, were considered as non-essential business and had to close their doors and find new ways to keep business afloat (Goolsbee et al., 2020).

Using a case study approach, this chapter explores how some brands and non-essential businesses adapted their marketing strategies and offerings to comply with the COVID-19 regulations. The first section presents a literature review related to business strategies, strategic flexibility, and consumer behavior during a crisis. Followed by the research method, and the analysis of marketing strategies of 10 companies or brands. The chapter ends with a discussion of the findings, limitations, and suggestions for future research.

3.2 LITERATURE REVIEW

3.2.1 EFFECTS OF THE COVID-19 PANDEMIC ON BUSINESS STRATEGIES

At the beginning of the lockdown, authorities classified businesses into essential and non-essential (Goolsbee et al., 2020). As the lockdown went by and regulations and restrictions became the norm, firms' options were clear: innovate to face the challenge or die.

As in any other crisis, the COVID-19 presented high levels of uncertainty (Grewal and Tansujah, 2001), and strategic flexibility and market orientation became a must (Eggers, 2020). Small firms faced challenging decisions: with fewer resources and fewer employees, they became more vulnerable to the lockdown measures (Eggers, 2020).

Suddenly managers had to find new ways to operate their business and optimally service their customers during the crisis. While some simply reacted by changing store hours and adding safety measures to keep both employees and consumers safe, many firms developed new capabilities that helped them survive during the crisis and provided them with strategic tools to be employed in the future.

And although before the COVID-19 crisis, consumers had already begun to move toward online shopping, during the lockdown period, individuals seeking convenience and safety changed their shopping behavior dramatically, pushing firms to adapt to the challenge of a volatile, uncertain, complex, and ambiguous (VUCA) world, where anxiety, fear, and fight, kill, and stress became the norm (Bennett and Lemoine, 2014). As Taleb (2010) suggests, a Black Swan was born with COVID-19. The disease that quickly spread from one continent to another came as a complete surprise for everybody, requiring proactiveness and organizational agility at all firms' levels to maintain their competitiveness (Al-Omoush et al., 2020) and remain relevant for their customers.

While many firms had no other choice than to react or respond to the COVID-19 environment, proactive firms introduced new products and services and innovated their marketing strategies, diversifying their offerings and acquiring more potential customers (Al-Omoush et al., 2020; Wang et al., 2020). Darks stores and ghost kitchens came to the rescue of traditional business during the pandemic (Arrojado, 2021; Bitterman and Hess, 2020; Verma, 2015), becoming an essential link of the delivery chain of restaurants and supermarkets. A dark store resembles a supermarket or a warehouse, but it is not open to the general public. Dark stores are open only to delivery pickers (e.g., Instacart, Rappi, etc.), who come to the store to fill pre-placed digital orders and are not interested in fancy displays or other amenities, allowing owners of the store to increase their profitability per square foot, while still meeting their customers' needs and demands (Verma, 2015).

On the other hand, ghost kitchens are rented workspaces allowing restaurant owners to have a shared kitchen at a strategic location, where cooking is done only for takeout and delivery orders (Safian-Demers, 2020). This new dining model allows restaurants to optimize their operations and serve fearful customers, who want to avoid long lines to enter the restaurant or be served at the counter but still want to enjoy their favorite meals from the safety of their household.

As Cromwell and Kotelli (2021) suggest, during the pandemic, some firms employed a traditional innovation process to satisfy their current customers, with a modified strategy (e.g., curbside pickup), while the most disruptive ones made use of breakthrough technologies to attract new market segments. Cromwell and Kotelli (2021) developed a framework to assess their innovation capabilities based on how firms responded to the pandemic. This framework classifies firms in four different categories:

1. **Trailblazers:** Companies that used their existing resources, knowledge, and capabilities to help solve a COVID-19 problem. For example, clothing brands selling face masks.
2. **Trendsetters:** Firms developing best-in-class solutions to cover the COVID-19 demand but simultaneously seeking to create sustainable strategies for the future. That includes hospitals offering telemedicine or companies developing new software for remote work.
3. **Adapters:** Companies with a temporary strategy to maintain the demand, like restaurants opening spaces for outdoor dining.
4. **Optimizers:** Sustainable strategies built to cover current demand by increasing their efficiency to serve customers. For example, hardware stores like Home Depot using augmented reality to sell more products; banks using QR codes so customers can get an appointment to go to the bank instead of waiting outside in a queue.

And although most firms around the World were creative and tried to reduce costs to keep their doors open, one of their biggest challenges was to increase- or at least maintain – the demand from those fearful and conservative consumers (Kantar, 2020), who alarmed by the COVID-19 changed their behaviors as we will observe in the next section.

3.2.2 FEAR, BOREDOM, AND ANXIETY: TRIGGERS OF NEW SHOPPING BEHAVIORS

Natural disasters (e.g., hurricanes, tornadoes, earthquakes, etc.), and economic and political crises usually affect consumers' behaviors. Previous research suggests that individuals who live a natural disaster, for example, tend to modify their shopping motives before, during, and after the event: alerted by the news, they stockpile basic supplies to be able to survive during the event, and sometimes they even get different products to defy boredom during the event. Once the storm is over, they replace the damaged items and acquire goods to celebrate the end of the catastrophic event (Zwanka and Buff, 2020). But, what happens during a complex and lengthy situation, such as the COVID-19 pandemic?

To cope with a crisis, most individuals need to understand what is happening, feel safe, and have some sense of control of the situation (Barnes et al., 2020). Therefore, irrational and hoarding behaviors, like stockpiling cleaning supplies, toilet paper, bottled water, or binge-watching the news, emerging among consumers during a crisis (Kirk and Rifkin, 2020; Laato et al., 2020). During the COVID-19 lockdown, consumers' unusual shopping behavior was also exacerbated by what they saw in the news or in social networks: empty shelves at the stores, aggressive behaviors exhibited by shoppers fighting for items, and new retail policies, that limited the purchase of certain products, leaving individuals with a sense of scarcity (Kirk and Rifkin, 2020); with a feeling of "caught off guard" (Zwanka and Buff, 2020) and a significant need to have enough information to make informed decisions (Laato et al., 2020).

Forced to be at home, soon individuals began new daily rituals, like cooking from scratch, planting a garden or even, redecorating the house to adapt it to the unique needs of the family (Gammon, 2020; Kantar, 2020), which allowed members of the household to have a resource to cope with the lockdown, replace the fear and instill a sense of normality (Imber-Black, 2020). Parents had to become creative and find new ways to celebrate birthdays, anniversaries, and holidays (Whiten, 2020), board games and toys provided a way to strengthen family bonds, video conferences allowed grandparents to share time with their grandchildren, and virtual reality allowed people to experience a different form of traveling (Zwanka and Buff, 2020).

Given that during the COVID-19 pandemic, authorities also implemented social distancing measures and limited social encounters, the situation became even more painful and stressful for most individuals (Kirk and Rifkin, 2020), who then turned to their favorite brands seeking for some sort of social connectedness (Reimann et al., 2017). Additional to the fear and the stress, it was also observed that boredom and anxiety soon installed in most households, and parents had to find resources to keep both their children and themselves busy, and feeling secure. Previous research suggests that everyday creative activities enhance people well-being and makes them feel better emotionally (Conner, De Young, and Silvia, 2018; Karwowski et al., 2021); it also suggests that individuals who feel isolated or anxious seek particular material objects (e.g., a favorite toy, a cozy sweatshirt, etc.), to soothe their distress (Reimann et al., 2017), which provided brands both with a challenge and an opportunity to be present, delight their customers, and help them cope with the pandemic (Barnes et al., 2020).

Qualitative research was conducted to identify the strategies employed by different firms to adapt to these new shopping behavior patterns. The methodology is explained in further sections.

3.3 METHODOLOGY

Explorative qualitative research was used to gain a deeper understanding of how different firms adapted their offerings and marketing strategies during the COVID-19 lockdown. For this research project, the case study approach was employed (Bonoma, 1985). This approach is appropriate for obtaining details and understandings about a specific phenomenon under study. The case study approach allows researchers in the area of marketing to observe the phenomenon within its natural settings. To get a fuller picture of the phenomenon, case study relies on personal interviews, unobtrusive observations, and secondary data. Although there are no rules on the number of cases to be included in the case study approach, a common knowledge rule suggests to include a minimum of two and a limit of 12 (Perry, 1998).

A purposeful sample of 10 organizations was selected. The criteria to select the sample was using an innovative business strategy to help consumers and clients survive and thrive during the COVID-19 pandemic.

Seven of the chosen organizations are well-known international firms, and their novel strategies were mentioned in at least one specialized marketing and business news source. And the other three are Mexican enterprises that used Instagram and Facebook to inform the public about their new services during the lockdown.

To collect data, the author consulted companies' websites, news distribution sites (e.g., PR Newswire, Marketing Dive and MediaPost Publications) and conducted interviews with the owners of two of the studied business. The exploration and the interviews allowed the author to identify best practices in marketing and make some generalizations, as explained in further sections.

3.4 RESULTS

3.4.1 THE FIRST ONE TO MOVE IS THE ONE THAT REACHES MORE CONSUMERS

To survive the crisis, firms had to seek new market opportunities. And to be able to serve their fearful and anxious customers, most had to re-design their offerings, re-imagine the customer experience and find new ways to deliver positive service encounters. Even priests and cult ministers had to get creative to offer parishioners their religious services and provided the much-needed spiritual and religious comfort that some people seek during the lockdown.

While some educational services, like elementary school, high school, or college, education, clearly remained relevant for most families who quickly engaged in a homeschooling program, other forms of extracurricular and educational services (e.g., Ballet lessons, art classes, soccer practice, etc.), were significantly affected by the COVID-19 restrictions, leaving instructors and owners of studios and gyms with the need to find alternative methods of delivery to reduce physical proximity, having to rely on digital media and other tools to reach their audience and stay open.

Soon social networks and video conference platforms like Zoom became the option to have social, educational, work-related, and religious encounters. But even when digital technologies allowed to provide different services (i.e., education, entertainment, conferences, religious preaches, etc.), something was lacking. Children needed supplies for their school projects, parents required ingredients to cook and bake, and youngsters soon became bored. At the same time, firms had a stock of goods that needed to reach the market, at least to recover part of their investments, and be able to pay the bills.

What did firms do to comply with the new demands of their customers? What strategies did they employ to keep business afloat? Our study develops some exciting findings. Trying to reduce long queues and reduce physical contact, banks began to rely on QR codes to schedule appointments for their clients. In contrast, retailers began to offer online ordering, direct delivery, and curbside pick-up services, trying to make the shopping experience as seamless as possible. At the same time, some non-essential businesses had to re-shape their offerings if they wanted to remain relevant for consumers.

Using all their creativity, firms designed new marketing strategies and created alternative means to deliver their offerings to keep consumers interested and their wallets wide open. Table 3.1 presents some examples of the new services and the promotional strategies employed by the sampled firms in this study.

TABLE 3.1 Examples of Marketing and Delivery Strategies Employed by Some Firms

Company	Main Product/Service	Strategy During COVID-19 Lockdown
Bud Light	Beers	**Bud Light Console:** Limited edition of a gaming console, shaped like a 6-pack of beers, loaded with six video games. The device was sold by auction on the brand's dedicated microsite.
Campbell's	Multi-national food company	**Save the Snow Day Kit:** Free limited-edition snow day activity kits sent to individuals that signed a pledge on a dedicated microsite.
Coca Cola Company	Non-alcoholic beverage company	**My Safe Store:** Kit for small retailers, including gloves, face masks, antibacterial gel, and protective screens.
Mexican Floral Technical Institute	School of floral design	**Boxed Flowers Workshops:** Kits to make floral ornaments at home, with live Zoom sessions to guide participants.
Lauren Peacock	Book author; creator of subscription boxes for divorced women.	**The Divorce Case:** A monthly subscription box, including seven different items, such as chocolates, toiletries, and beauty supplies. As well as a membership to become part of a social network community.

TABLE 3.1 *(Continued)*

Company	Main Product/Service	Strategy During COVID-19 Lockdown
Marce Art Gallery	Independent artist and art gallery.	**Wine and Canvas:** A kit including a blank canvas, brushes, paints, and a bottle of wine.
Mimo	Early child stimulation school, based in the North of Mexico.	**Mimo in a Box:** A seasonal kit containing educational materials for parents to play and engage with children under five years. Videos with additional information and activities were sent via WhatsApp to those who got the kit. Kits were released every quarter: Easter, Summer, Fall, and Christmas time.
Philadelphia	Cream cheese.	**Philly Series 5:** A baking kit including a baking pan, Philadelphia cream cheese, sugar, vanilla extract, chocolate sandwich cookies, white chocolate, wax paper, and instructions to bake a cake (https://phillyseries5.com/)
Qingdao Publicity Department	Qingdao is a city in China seeking to attract visitors.	**Gift from Qingdao:** 10,000 mystery boxes were distributed to tourists in Qingdao, including postcards, tickets for a local museum, a book about the city's culture, and a voucher to try Qingdao's food specialties at a local restaurant.
Trident	Chewing gum.	**Trident Chew Tone:** Limited edition packages including a QR code to unlock interactive AR experiences and the opportunity to participate in a sweepstake and win tickets to the Grammy Awards.

Source: Own elaboration.

Following Cromwell and Kotelly's (2021) framework, we also classified the studied firms and identified some interesting strategies that allowed firms to remain relevant for their consumers. Table 3.2 presents the classification of the studied firms.

3.5 DISCUSSION AND CONCLUSIONS

It is important to acknowledge that the consumer reality has changed dramatically over the COVID-19 pandemic. Both goods and services firms need to take advantage of consumers' new shopping habits and behavior and be prepared to help them face the risks that this virus represents. With a new environment and new rules, most companies have been forced to innovate and become more creative.

TABLE 3.2 Firms' Classification

Trailblazer	**Trendsetter**
Qingdao Publicity Department	Mimo
	Mexican Floral Technical Institute
	Marce Art Gallery
	Lauren Peacock's Divorce case
Adapter	**Optimizer**
Bud Light	Coca Cola bottling company's safe retail kit
Campbell's	
Philadelphia	
Trident	

Source: Own elaboration, based on Cromwell and Kotelly (2020).

We discovered that to face the COVID-19 business crisis, there was no "one size fits all" solution, but common ingredients were present such as creativity, flexible thought, and use of digital tools. Firms that quickly adapted and provided a rapid response to the changing demands are the ones that have a better chance of surviving for long-term success in this complex reality. Given that individuals have changed their purchase priorities, are more aware of the risks of the virus, managers, entrepreneurs, and employees need to be flexible and eager to provide not only the goods and services demanded by consumers but to modify their business models to remain relevant for those individuals seeking for safety, not only with the items they buy but with the way online transactions are done paying particular attention to cybersecurity.

Our findings suggest that some strategies, like the Bud Light Console, even served as a tool to demonstrate empathy and solidarity with those who lost their jobs during the lockdown. Proceeds from the sale of the beer gear went to the program "Change is on the Menu" from the National Restaurant Association to help restaurant workers displaced by the pandemic. Such strategies denote how firms are willing to sell a product, make a profit, and become part of the solution to improve the well-being of their stakeholders.

Dedicated microsites were launched to invite consumers to get involved with the brand and the promotions, allowing firms to engage with the audience, gather data from those who participated in the sweepstakes, and share the firm's values and genuine interest for the community.

It was clearly demonstrated that the online World and e-commerce are a go-to destination for many brands. Digital technologies and new forms of delivering goods and services provided competitive advantages to firms and helped them survive and improve their business models. Tutorials on YouTube, TikTok, and Instagram were welcomed by individuals seeking to acquire new pieces of knowledge, exercise at home, or embrace a new hobby as an antidote to loneliness and boredom during COVID-19. At the same time, downloadable, and printable activities designed by several brands became a valued resource for busy parents who needed to keep their kids busy.

The findings also suggest that those firms and entrepreneurs that faced their fears and decided to use their resources and creativity to implement new strategies are the ones that not only survived the crisis but expanded their markets with novel offerings. Aided with digital technology, small firms took a leap of faith and moved their face-to-face offerings to the digital environment, finding eager consumers not only in their actual markets but even in different regions of the country, and in some cases, in foreign cities. As Akpan et al. (2020) suggest, cutting-edge technologies helped businesses acquire competitive advantages during COVID-19. Digitization, internet of things (IoT), artificial intelligence (AI), and virtual reality enabled new business models, offering safety and convenience to shoppers 24 hours per day, and seven days per week.

At the same time, it is important to acknowledge that the COVID-19 crisis will have an impact on the built environment and the viability of shopping districts and other retail destinations. It seems that a new era of shopping has begun. As more shoppers get comfortable with the idea of buying online, visiting traditional stores or shopping malls might not be as popular as it used to be before the pandemic arise; leaving shopping venues with a significant challenge in their hands: how to attract individuals to visit, explore, and hang out at these locations?

3.6 LIMITATIONS AND FUTURE RESEARCH

Some questions remain unanswered. Although the studied firms did sell their boxes, do consumers trust the boxing process? Or are firms using some sort of safety seal to provide individuals with a guarantee that the items in the box were adequately sanitized before being included in each kit?

The findings of this research are based on the researcher's interpretation of the observed marketing strategies used by the firms. To extend the findings to different businesses, triangulation is needed and the use of financial data to measure the success of the strategies.

A year has passed since the lockdown began; therefore, future studies are needed to assess consumers behavior and identify if once the authorities relaxed the lockdown measures and allowed businesses to reopen their facilities, shoppers kept the habits acquired during the COVID-19 quarantine or went back to what they call the old normality. Or if the trends that started in March 2020 become the new normal (i.e., working, studying, and exercising at home), firms need to change the way they do business in the long run, and of course, further research is required to assess how consumers adapt and respond to these new business strategies.

KEYWORDS

- **COVID-19**
- **curbside pick up**
- **home delivery**
- **marketing strategies**
- **subscription boxes**
- **unboxing**

REFERENCES

Akpan, I. J., Soopramanien, D., & Kwak. D. H., (2020). Cutting-edge technologies for small business and innovation in the era of COVID-19 global health pandemic. *Journal of Small Business & Entrepreneurship,* 1–11.

Al-Omoush, K. S., Simón-Moya, V., & Sendra-García. J., (2020). The impact of social capital and collaborative knowledge creation on e-business proactiveness and organizational agility in responding to the COVID-19 crisis. *Journal of Innovation & Knowledge, 5*(4), 279–288.

Arrojado, C., (2021). *"Why 'Host Kitchens' Are the Next Big COVID Trend".* QS Magazine. Retrieved from: https://vtechworks.lib.vt.edu/bitstream/handle/10919/102983/Why%20%27Host%20Kitchens%27%20are%20the%20Next%20Big%20COVID%20Trend%20_%20QSR%20magazine.pdf?sequence=1&isAllowed=y (accessed on 18 February 2022).

Baar, A., (2020). *Bud Light Taps into Gaming Wars with 6 Pack-Console.* Retrieved from: https://www.marketingdive.com/news/bud-light-taps-into-gaming-wars-with-6-pack-console/589278/?utm_source=Sailthru&utm_medium=email&utm_campaign=Issue:%202020-11-18%20Marketing%20Dive%20Newsletter%20%5Bissue:30944%5D&utm_term=Marketing%20Dive (accessed on 18 February 2022).

Barnes, D. C., Mesmer-Magnus, J., Krallman, A., Scribner, L. L., & Guidice. R. M., (2020). Customer delight during a crisis: Understanding delight through the lens of transformative service research. *Journal of Service Management, 32*(1), 129–141.

Barnes, S. J., Diaz, M., & Arnaboldi. M., (2021). Understanding panic buying during COVID-19: A text analytics approach. *Expert Systems with Applications, 169*, 114360.

Bennett, N., & Lemoine. J., (2014). *What VUCA Really Means for You* (Vol. 92, No. 1, 2). Harvard business review.

Bitterman, A., & Hess. D. B., (2020). *Going Dark: The Post-Pandemic Transformation of the Metropolitan Retail Landscape.* Retrieved from: http://cp-cloudpublish-public.s3.amazonaws.com/p6/5f92d2f414631.pdf (accessed on 18 February 2022).

Bonoma, T. V., (1985). Case research in marketing: Opportunities, problems, and a process. *Journal of Marketing Research, 22*(2), 199–208.

Campbell's, (2020). *Campbell's Soup Company.* Retrieved from: https://www.campbellsoupcompany.com/about-campbell/ (accessed on 18 February 2022).

CDMX, (2020). Gobierno de México. *Reactivación de la Economía Mexicana de Manera Responsable y Segura.* Retrieved from: https://www.gob.mx/covid19medidaseconomicas/acciones-y-programas/nueva-normalidad-244196 (accessed on 18 February 2022).

Coca Cola, (2020). *Hagamos Esto Juntos: Trabajamos Juntos Por Los Negocios Locales.* Retrieved from: https://www.coca-colamexico.com.mx/noticias/comunidad/juntos-por-los-negocios-locales (accessed on 18 February 2022).

Conner, T. S., DeYoung, C. G., & Silvia. P. J., (2018). Everyday creative activity as a path to flourishing. *The Journal of Positive Psychology, 13*(2), 181–189.

Cromwell, J., & Kotelly. B., (2021). *A Framework for Innovation in the COVID-19 Era and Beyond.* MIT Sloan management review. Retrieved from: https://sloanreview.mit.edu/article/a-framework-for-innovation-in-the-covid-19-era-and-beyond/ (accessed on 18 February 2022).

Eggers, F., (2020). Masters of disasters? Challenges and opportunities for SMEs in times of crisis. *Journal of Business Research, 116*, 199–208.

Gammon, K., (2020). *Kneading to Relax? How Coronavirus Prompted a Surge in Stress Baking.* The Guardian. Retrieved from: https://www.theguardian.com/us-news/2020/apr/19/coronavirus-stress-baking-sourdough-kneading-relax?utm_campaign=Sophie%20IotD&utm_source=hs_email&utm_medium=email&utm_content=87464861&_

hsenc=p2ANqtz-9zMbbf-HpUy9Z_3iVYS1QJpMmshsdoU-CSFbW7Rs1UEg76_9s_2 jtxNTNN25Ko9gyY2DEPHbCuMRpb2boCRwEqoojk4g&_hsmi=87464916 (accessed on 18 February 2022).

Garfin, D. R., Silver, R. C., & Holman. E. A., (2020). The novel coronavirus (COVID-2019) outbreak: Amplification of public health consequences by media exposure. *Health Psychology, 39*(5), 355–357-

Goolsbee, A., Luo, N. B., Nesbitt, R., & Syverson. C., (2020). *COVID-19 Lockdown Policies at the State and Local Level*. University of Chicago, Becker Friedman Institute for Economics Working Paper, (2020–116).

Grewal, R., & Tansuhaj. P., (2001). Building organizational capabilities for managing economic crisis: The role of market orientation and strategic flexibility. *Journal of Marketing, 65*(2), 67–80.

https://www.prnewswire.com/news-releases/campbells-soup-unveils-savethesnowday-to-preserve-day-of-play-for-kids-across-america-301183207.html (accessed on 18 February 2022).

Imber-Black, E., (2020). Rituals in the time of COVID-19: Imagination, responsiveness, and the human spirit. *Family Process, 59*(3), 912–921.

IMFT, (2020). Mexican Floral Technical Institute. *Our Courses*. https://www.imtf.com.mx/cursos (accessed on 18 February 2022).

Kantar, (2020). *10 Thoughts on COVID-19 and its Impact on Purchase and Consumption Behavior*. Retrieved from: https://www.kantar.com/inspiration/coronavirus/10-thoughts-on-covid-19-and-its-impact-on-purchase-and-consumption-behaviour (accessed on 18 February 2022).

Karwowski, M., Zielińska, A., Jankowska, D. M., Strutyńska, E., Omelańczuk, I., & Lebuda. I., (2021). Creative lockdown? A daily diary study of creative activity during pandemics. *Frontiers in Psychology, 12*, 23.

Kirk, C. P., & Rifkin. L. S., (2020). I'll trade your diamonds for toilet paper: Consumer reacting, coping, and adapting behaviors in the COVID-19 pandemic. *Journal of Business Research, 117*, 124–131.

Laato, S., Islam, A. N., Farooq, A., & Dhir. A., (2020). Unusual purchasing behavior during the early stages of the COVID-19 pandemic: The stimulus-organism-response approach. *Journal of Retailing and Consumer Services, 57*, 102224.

Marketing Charts, (2020). *4 Points About Consumer Behavior Changes Spurred by COVID*. https://www.marketingcharts.com/featured-115684 (accessed on 18 February 2022).

Marketing Dive, (2020). *Campbell's Soup Pushes to Save Snow Days from Virtual Classes*. Retrieved from: https://www.marketingdive.com/news/campbells-soup-pushes-to-save-snow-days-from-virtual-classes/591627/?utm_source=Sailthru&utm_medium=email&utm_campaign=Issue:%202020-12-04%20Marketing%20Dive%20Newsletter%20%5Bissue:31240%5D&utm_term=Marketing%20Dive (accessed on 18 February 2022).

Mathur, A., Moschis, G. P., & Lee. E., (2003). Life events and brand preference changes. *Journal of Consumer Behavior: An International Research Review, 3*(2), 129–141.

Mondelez International, Inc., (2021). https://www.prnewswire.com/news-releases/trident-gum-and-t-pain-team-up-to-launch-new-limited-edition-trident-gum-chew-tunes-packs-301094916.html (accessed on 18 February 2022).

Pantano, E., Pizzi, G., Scarpi, D., & Dennis. C., (2020). Competing during a pandemic? Retailers' ups and downs during the COVID-19 outbreak. *Journal of Business Research, 116*, 209–213.

Peacock, L., (2021). *In Advance of Valentine's Day, "The Divorce case" Subscription Box Offers Community, Self-Care and Laughs to Divorcees Everywhere*. Retrieved from: https://www.prnewswire.com/news-releases/in-advance-of-valentines-day-the-divorce-case-subscription-box-offers-community-self-care-and-laughs-to-divorcees-everywhere-301205855.html (accessed on 18 February 2022).

Perry, C., (1998). Processes of a case study methodology for postgraduate research in marketing. *European Journal of Marketing, 32*(9), 785–802.

Perry, C., (2000). Case research in marketing. *The Marketing Review, 1*(3), 303–323.

PWC, (2020). *COVID-19: Oportunidades Para los Mercados de Consumo*. Retrieved from: https://www.pwc.com/mx/es/gestion-de-crisis/covid-19/covid-19-oportunidades-mercados-consumo.html (accessed on 18 February 2022).

Reimann, M., Nuñez, S., & Castaño. R., (2017). Brand-aid. *Journal of Consumer Research, 44*(3), 673–691.

Rodríguez, C. M., (2020). *Wine and Canvas Saltillo México*. https://www.instagram.com/wineandcanvas.mx/?hl=es-la (accessed on 18 February 2022).

Safian-Demers, E., (2020). Ghost kitchens. *Wunderman Thompson Trend Reports*. https://intelligence.wundermanthompson.com/2020/12/ghost-kitchens/?mc_cid=8ebffdf279&mc_eid=[34b846a654] (accessed on 18 February 2022).

Save the Snow Day, (2020). https://www.savethesnowday.com/ (accessed on 18 February 2022).

Seetharaman, P., (2020). Business models shifts: Impact of COVID-19. *International Journal of Information Management, 54*, 102173.

Sheth, J., (2020). Impact of COVID-19 on consumer behavior: Will the old habits return or die?. *Journal of Business Research, 117*, 280–283.

Stadt Quingdao, (2021). *The "Gift from Quingdao" Program is Expected to Become an Annual Campaign in Quingdao in the Future*. Retrieved from: https://www.prnewswire.com/news-releases/the-gift-from-qingdao-program-is-expected-to-become-an-annual-campaign-in-qingdao-in-the-future-301201085.html (accessed on 18 February 2022).

Taleb, N. N., (2010). *The Black Swan: The Impact of the Highly Improbable* (2nd edn.). Random House Publishing.

Verma, E. S., (2015). *Dark Stores, a Bright Concept in Franchising*. Retrieved from: https://www.franchiseindia.com/content/dark-stores-a-bright-concept-in-franchising.5965#:~:text=Dark%20stores%20have%20numerous%20benefits,stores%20can%20do%20deeper%20integrations (accessed on 18 February 2022).

Wang, Y., Hong, A., Li, X., & Gao. J., (2020). Marketing innovations during a global crisis: A study of China firms' response to COVID-19. *Journal of Business Research, 116*, 214–220.

Whiten, S., (2020). *Parents Look to Create a Magical Easter*. Retrieved from: https://www.cnbc.com/2020/04/02/parents-look-to-create-a-magical-easter-despite-coronavirus-restrictions.html (accessed on 14 March 2022).

Zwanka, R. J., & Buff. C., (2021). COVID-19 generation: A conceptual framework of the consumer behavioral shifts to be caused by the COVID-19 pandemic. *Journal of International Consumer Marketing, 33*(1), 58–67.

CHAPTER 4

Influencer Marketing in an Uncertain Economic Climate: A Wasted Opportunity

ISABEL MORTEO

Faculty of Tourism and Marketing, Autonomous University of Baja California, Tijuana, Mexico, E-mail: imorteo@uabc.edu.mx

ABSTRACT

Understanding the degree of responsibility that an individual who is regarded as relevant to the public holds towards their audience is often downplayed. Being aware of this responsibility would most certainly cause most individuals to question themselves twice before sharing their opinion on a specific matter. In the case of crisis management and decision making, an individual's statement could have unforeseen consequences. The COVID-19 pandemic has put forward the complexity of managing multiple sources of information by individuals, who in a sea of data and opinions have a hard time dissecting between truth, rumor, suggestion, and mandate, since the emphasis on the best course of action is not clearly pointed out for them. This chapter focuses on the relevance of being aware of the consequences of such inconsistencies in terms of how a well thought out influencer marketing strategy could benefit brands in the process of selecting individuals who will shed a positive light on their products and not end up behaving in questionable ways which end up forcing brands to sever their relationships with them.

Marketing by Contingency in the Time of COVID-19: Overcoming Business Crises and Meeting Marketing Challenges. Alicia De La Peña, PhD & Juan Bernardo Amezcua Nunez, PhD (Eds.)

4.1 INTRODUCTION

When discussing the topic of influencers, a specific image comes to mind: beautiful people benefiting from their leverage with companies in order to enjoy outstanding experiences, having access to exclusive products that mostly remain elusive to regular consumers. On the other hand, there is another kind of influencer: individuals who look like regular people, live like regular people (but better) and that one can relate to (to some extent) and that make you wish you had that kind of life – which seems realistic and achievable.

But how do these individuals become protagonists in this crowded world? This question has been extensively researched, including the requirements they have to meet in order to solidify their relationship to their audience (Jin, Muqaddam, and Ryu, 2019), the mechanisms through which they establish collaborations with brands (Delisle and Parmentier, 2016; Morteo, 2017), how their work provides them with an income source (Martínez-López et al., 2020; Morteo, 2017) and the impact the communication has on their audience and how they transform prospects into clients for the brands with which they are collaborating (Bowman, 2020; Jin et al., 2019).

Previous research on the topic has not been exhausted – their relevance as protagonists in marketing strategies has nowhere near reached its peak and is constantly evolving, along with the media through which they establish their relationship with consumers. Changes in format and interactions available in the myriad of social media platforms they use happen at an extraordinary speed, which forces both content creators and consumers to change and adapt with them, thus keeping everyone on their toes (Martínez-López et al., 2020).

Now, as if this constantly evolving digital environment was not enough, the year 2020 proved to be as turbulent and unpredictable as ever. Late 2019 (World Health Organization, 2020a) an unknown virus, identified first as cases of pneumonia and later as a new strain of Coronavirus, unleashed a pandemic with global reach, which brought the world to a halt in a way no one could have foreseen. By March 2020 (WHO, 2020b) the epicenter of the pandemic had shifted to Europe, and actions towards slowing the spread began taking shape in the form of restrictive travel, school closures, and businesses closing down due to the unfamiliar behavior of the virus.

The consequences of this pandemic had an enormous impact on businesses, especially when it comes to their marketing strategies. One of the most impactful initiatives put in place in order to slow the spread of COVID-19, as was this strain later named (Centers for Disease Control and Prevention, 2020a), was a period of quarantine (Centers for Disease Control and Prevention, 2020b) which prevented individuals from carrying out their regular activities such as going to the grocery store (Zhu, Xie, and Veeck, 2020), participating in events like concerts or even going out to restaurants. Additionally, the relevance of the topic has prompted the emergence of independent initiatives such as The COVID Tracking Project (The Atlantic Monthly Group, 2020) which have taken into their own hands gathering information that will aid the public to learn about the pandemic, its evolution, recommendations, and latest developments. The purpose of this chapter is to dissect behaviors of individuals identified as influencers relevant to the COVID-19 pandemic and how their audience responded to them, and what brands can learn from them in order to adapt their influencer strategy to the current scenario in the midst of an uncertain and unstable economy.

4.2 INFLUENCERS AS BRAND'S STRATEGIC PARTNERS

The appeal of influencers as part of a successful marketing strategy in collaboration with brands comes in the form of their ability to share a message intended for the influencer's audience. The format of that message can either be requested by the organization or consist of a set of expected pointers and some degree of freedom to create a message (Morteo, 2017), which conveys the authenticity of opinion that consumers expect from such individuals. Ultimately, the success of this collaboration is seen in the level of acceptance and agreement manifested by the influencer's audience, which is expected to be translated into increased sales.

This brief account of the dynamic of mutually beneficial collaboration established between brands and influencers will serve as a backdrop to explore their performance during the COVID-19 pandemic. Since the definition of influencer can be oversimplified to prominent figures that appeal to a wide audience, this concept encompasses individuals whose notoriety is due solely because of their activities linked to social media, as well as any other profile that captures their audiences' attention, regardless of the industry in which they are perceived as knowledgeable (Delisle and Parmentier, 2016), whether or not there is any support to their claims. It is this specific detail which begs the question: should we be paying this much attention to just about anyone? How should we discriminate between an individual who is worth listening to and someone who is not? Are there specific criteria that should help us discriminate who we are paying attention to?

4.2.1 HOW DO INFLUENCERS COLLABORATE WITH ORGANIZATIONS?

Organizations establish a relationship with their current and intended customers and a wider audience through the messages they share through multiple channels. In order for an organization's communication efforts to perform effectively, the messages shared are expected to positively contribute towards corporate objectives. Of course, there are challenges in designing such messages and managing multiple channels simultaneously, as shown in the study by Laurie and Mortimer (2019), who explored the importance of establishing a productive relationship between advertising agencies and clients. The main issue raised in their study was how organizations collaborate with advertising agencies in terms of information sharing and trust, specially emphasizing the challenge that integration of all communication channels represent, highlighting the complexities of achieving a true synergistic dynamic between multiplicity of actors.

Shifting our focus to the consumer, who is the target of simultaneous stimuli through a multiplicity of channels, their attention is a company's main interest. Finne and Grönroos (2017) provide a framework to analyze how the customer perceives, analyzes, and makes decisions based on the information they receive. An important shift in their conceptualization is that of to stop overlooking the true value that the customer ultimately confers to these messages, stressing the importance of utility for the customer, otherwise risking becoming irrelevant for decision-making purposes.

The trigger that gave influencers a prominent role as part of a marketing plan was the possibility of users to suppress unwanted advertisements (Martínez-López et al., 2020), which made companies look for alternatives to reach potential customers. Influencers became the logical choice since they are assumed to have leverage towards their audience, starting with the mechanism in which they acquire their status: content creation which appeals to individuals who choose to value their opinion, especially if they can agree with it (Chopra et al., 2020). If we were to establish the main criteria between celebrities and influencers, which would seem to perform quite similarly, it would be how they acquire their prominence, which for the case of celebrities is through their work and visibility in traditional media, whereas an influencer emerges due to the engagement of an audience that values the content they create and share through their social media accounts (Jin, Muqaddam, and Ryu, 2019).

The freedom with which these collaborations are carried out has evolved. The prominence of such messages in the eyes of the consumer has prompted the creation of guidelines to regulate them, such as disclosure requirements by the social media platforms where these messages are shared, including mentions of the degree involvement of the brand with the influencer (Martínez-López et al., 2020; Weismueller et al., 2020). There have also been consequences for these individuals, who as a tradeoff for their popularity are exposed to slander and defamation with no effective legal recourse for their protection (DeSimone, 2018). Lee (2018) succinctly summarized the pros and cons of collaborating with influencers highlighting its risks, among them the unpredictability of the influencers behavior which could end up damaging the brand. Despite the drawbacks, the appeal to collaborate with influencers does not seem to slow down, with expenditure for 2017 being $570 million dollars and estimates expected to reach in excess of $23 billion dollars by 2025 (Grand View Research, Inc., 2020).

4.2.2 CONSUMER DECISION-MAKING PROCESS AND WHERE INFLUENCERS CHANGE ITS COURSE

According to the marketing report by Hubspot for 2020 (Hubspot, 2020), content marketing would still be supported by influencer marketing activities, stressing the importance of connecting with audiences through niche-specific influencers identified as micro-influencers and noting that influencer posts are the fourth most important type of content that achieves consumer engagement with brand content, behind photo/imagery, video posts and text-based posts. But how are influencers activities regarded as actually having an impact on consumer behavior?

Stankevich (2017) conducted an extensive literature review on consumer buying behavior. Issues such as the evolution of consumer expectations and demands of an increasingly personalized offer are highlighted, as well as new marketing challenges that are to be addressed with a strategic focus, such as the convergence of simultaneous generations with differing needs, or cultural differences that stress differences among consumers depending on context and require adapting to its nuances. An interesting point raised by Stankevich (2017) is that of how consumers are exposed to heightened amounts of stimulus, which present a challenge for consumers in order to discriminate between them and catalog them as relevant or irrelevant depending on their own sets of beliefs (Moore et al., 2018).

Chopra et al. (2020) expand on the topic of consumer behavior and how influencers play a role in shaping it. One of the factors highlighted in their research was that of fit between the product and the influencer, since consumers expect their criterion-shaping informants to have some degree of expertise to support their advice. This will in consequence be of aid in building a relationship with the audience (Jin et al., 2019), by establishing points of opinion convergence which in turn increases their credibility. This dynamic could be described as self-nurturing, resumed as an individual with expertise is validated by an audience that trusts this expertise, although they might not have any evidence as to how trustworthy their opinions are, using criteria such as number of followers to adjudicate trustworthiness (Weissmueller et al., 2020).

Taking all these factors into consideration, influencers tend to the needs of consumers by offering information that is specifically tailored to their needs, through content that the audience deems as informative and relevant, helping them discriminate between a multitude of alternatives. One of the most important factors is that the audience needs to agree with the information that is provided to them, which will strengthen the bond that the audience has with the influencer. But what happens when an influencer behaves in problematic ways in the eyes of their audience?

4.3 PERFORMANCE OF INFLUENCERS DURING THE COVID-19 PANDEMIC: QUESTIONABLE BEHAVIORS AND HOW THEIR AUDIENCE REACTED TOWARDS THEM

As has previously been pointed out, influencers bear a responsibility both towards their audience and the organizations with which they collaborate. Keller (2016) provided a list of key communication objectives, including the creation of awareness and salience, building trust, eliciting emotion, inspire action, and connecting with people, purposes which can be met via the engagement of influencers into the marketing communication strategies performed by influencers (Ki, Cuevas, Chong, and Lim, 2020). The focus of this chapter, as stated before, is to analyze how individuals regarded as relevant to the conversation surrounding COVID-19 performed in the eyes of their audience.

One of such examples is that of TV personality Ellen DeGeneres, host of her own show "The Ellen DeGeneres Show." On April 6th, 2020, Ellen posted a video on YouTube, which was later removed and edited, called "Ellen Is Back, Sharing Words of Encouragement from Her Living Room." In it, the talk show host made remarks that equaled being quarantined with being in jail, venting her frustration for the restrictions imposed on civilian mobility. Although the comment was later reframed as a joke, its audience reacted negatively towards it, by highlighting the incompatibility with reality that Ellen's perception of "being imprisoned" really means (Lee, 2020).

Another case of ill-timed messaging was that of Gal Gadot, an actress whose interpretation of Wonder Woman in 2017 put her in the limelight (IMDb.com, n.d.). Back in March of 2020, the actress shared a video with a rendition of the popular song "Imagine," originally performed by The Beatles, where she was joined by a plethora of other actors joining in, each performing a line of the song. In his piece "This 'Imagine' cover is no heaven," Jon Caramanica critiques each appearance, focusing mainly on their appearance or lack of singing abilities, questioning what kind of contribution this performance could have given the pervasiveness of the pandemic (Caramanica, 2020). In her article titled "Gal Gadot Understands Why People Didn't Like Her "Imagine" Video," Julia Emmanuele (2020) dissects the incident and expands on its negative consequences, providing context to it by sharing that one of the actors participating in it, Jamie Dornan, affirms being aware of his own privileged conditions while recording his part, deciding on performing in a bathroom instead of sharing his true location, with oceanfront views. There are more severe cases, one of which is the focus of this chapter. Based on the previously reviewed literature, the research question that guides this chapter is: how does the audience react towards the behavior of an influencer that is deemed unacceptable in the context of the global COVID-19 pandemic?

4.4 METHODOLOGY

The choice of a qualitative research focus for this chapter rests on its purpose: to understand how individuals make sense of the reality they find themselves immersed into (Haradhan, 2018). In terms of the selection of the research methodology most appropriate with the phenomenon of interest, the case study was identified as most fitting since it requires intense scrutinization of either an individual or some other unit in order for the researcher to understand its variables and make sense of them with the goal of offering generalizations applicable to the phenomenon that is under study (Heale and Twycross, 2018). As for the analysis technique selected, Gaur and Kumar (2018) defined content analysis as a technique where the researcher structures information pertinent to a specific topic and the coding of the source material in order to draw conclusions. This technique can be applied either to perform a quantitative or qualitative analysis, where the main difference lies on whether its approach is deductive, where theory is the starting point to develop the coding scheme, or inductive, which requires the coding scheme to be developed inductively. For this chapter, the approach will be deductive, in order for the data to offer the relevant topics to discuss the phenomenon.

Bengtsson (2016) succinctly structured the path towards performing content analysis in four steps: planning, data collection, data analyzing and creation of the report and result presentation. As can be expected, the most complex step is that of data analysis, which requires, depending on approach, the completion of four stages: decontextualization, recontextualization, categorization, and compilation. Bengtsson's diagram allows for the process to be understood and easily approached.

4.4.1 CASE SELECTION

Arielle Charnas is a female entrepreneur, known for her business ventures which include the launch of her own standalone line of apparel, called Something Navy. Her behavior during the pandemic was the topic of several news outlets which chose to track and expose every move and fed the audience a narrative that invited questioning and condemning her behavior, as well as involving the most notorious brand with which she had previously had a collaboration: Nordstrom. In terms of relevance, the number of search results for keywords associated directly with her and her company, which can be seen in Table 4.1, as well as the topic at hand, warrants for consideration. With regards to the relevance of the case, the Google Trends analysis on the previously detailed keyword sets indicated a worldwide interest in Arielle Charnas,' as the keyword search relevance indicates a level of 100, which is regarded as the highest level of interest (Google Inc., 2021) between the months of March and May of 2020.

TABLE 4.1 Google Search Results per Keyword Set

Keyword Set	Results
Arielle Charnas	159,000
Something navy	253,000
Arielle Charnas COVID drama	92,900

Source: Elaborated by the author.

In addition to the relevance reflected by the Google Trends term search statistics, additional support for the importance of Arielle Charnas' as a public figure can be sustained by her social media presence, shown in Table 4.2. for both her personal and her brand's account. Even though Arielle has a following of over 1.3 million, tools can be used in order to assess the quality of her audience, which do not provide outstanding results, meaning her account behaves as other similar accounts, which can be seen in Table 4.3, resulting in an average audience quality score of 46 out of 100 (Influencer Marketing Hub, 2021).

TABLE 4.2 Instagram Account Main Statistics

Instagram Handle	Followers	Following	Posts
@ariellecharnas	1.3 M	962	8,261
@somethingnavy	297 k	28	1,871

Source: Elaborated by the author.

TABLE 4.3 Fake Follower Checker Report on Arielle Charnas' Personal Instagram Account

Instagram Handle	Followers	Engagement Rate	Likes–Comments Ratio
@ariellecharnas	1.3 M	2.04%	144.97 (good)

Source: Elaborated by the author with information from Influencer Marketing Hub (2021).

4.5 DATA ANALYSIS

Data collection was through a Google search performed for the 3 previously mentioned keywords: "Arielle Charnas," "something navy" and "Arielle Charnas covid drama". From each keyword set search, 10 articles were selected in order to define the sample of articles that were going to be analyzed following the principles of a content analysis so as to identify the main discussed topics and how diverse news outlets presented Arielle Charnas's behavior to the public. Since the keywords selected produced overlapping results, a total of 27 distinct articles comprises the final sample (Table 4.4).

TABLE 4.4 Coding Scheme for Content Analysis

Parent Category	Sub-Categories	Explanation
Focus	Fact reporting/ opinion piece	Whether the article focused on the narrative of events, or the author offered their interpretation or judgment on facts
Coronavirus mentions	–	Frequency with which the article uses the word coronavirus
Arielle's pandemic behavior	Yes/no	Whether the author includes facts on the influencer's decisions
Nordstrom partnership	Yes/no	Whether the author makes reference to the influencers' previous business association with Nordstrom
Covidiot	Yes/no	Whether the term "covidiot" is mentioned in the article
Nordstrom disassociation	Yes/no	Whether reference is made to Nordstrom's response with regards to a business association with the influencer
CDC guidelines	Yes/no	Whether the author includes reference to CDCs guidelines on behavior during the COVID-19 pandemic

Source: Elaborated by the author.

4.5.1 FINDINGS

Arielle Charnas's persona had been characterized by Crain's New York Business in their article "40 under 40 class of 2018" (Crain's New York

Business, n.d.) as a so called "everyday girl", in her own words. She had been found to be an individual who had amassed an impressive following due to the nature of her social media behavior: an account of her life which had resonated with more than 1 million individuals who were interested in her recommendations on topics as varied as restaurants, clothing, and even doctors. Salient conclusions due to content analysis:

- Out of the 27 articles, 20 mention her behavior, 14 being fact reporting (3 mentioning CDC guidelines) and 6 being opinion pieces where 2 make reference to CDC guidelines.
- Out of the 7 articles that don't reference Arielle's behavior during the pandemic and focus on her professional accomplishments and future endeavors, one makes a reference to her relationship to Nordstrom which had been ended by the time the article was written, and even includes the word "covidiot" in their characterization of her.
- Only 6 articles mention CDC guidelines as reference to contrast/ question appropriate behavior.
- Charnas is referred as "covidiot" in 6 articles, one of them announcing her pregnancy and another one focusing on the launch of her standalone line.
- 13 articles mention Charnas' relationship to Nordstrom: 8 of them delved into Charnas' questionable behavior during the pandemic – 6 of them made reference to Nordstrom disassociating themselves from her.
- 7 of the 27 articles of the selected sample that don't reference Arielle's behavior came from the keyword "something navy" – which might indicate a biased reporting on the matter.

Tangible evidence on the impact this situation had on Arielle's standing as an influencer can be appreciated in her follower count on her Instagram account, which is shown in Figures 4.1 and 4.2, where her account had an evident negative slope representing individuals who decided to stop following her account.

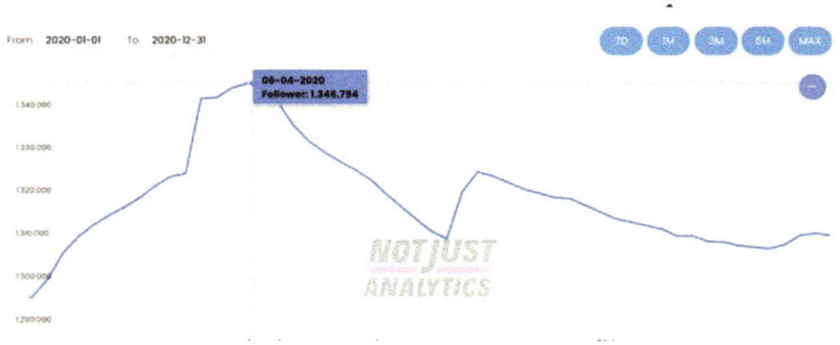

FIGURE 4.1 Arielle Charnas' follower count trend for April, 6th, 2020.

Source: Reprinted with permission from Not Just Analytics – Analyze Instagram Profiles Free (2021).

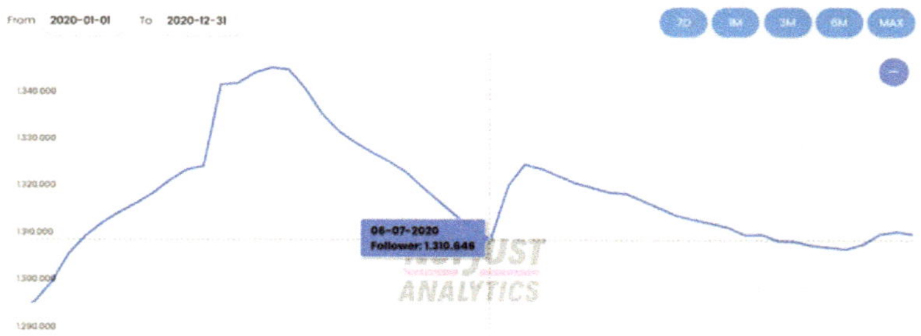

FIGURE 4.2 Arielle Charnas' follower count trend for July 6th, 2020.

Source: Reprinted with permission from Not Just Analytics – Analyze Instagram Profiles Free (2021).

4.6 DISCUSSION

This section will focus on highlighting managerial implications, limitations, and further research that is considered relevant to continue the understanding of this phenomenon. Managerial implications are specially explored since well-coordinated dynamics between brands and influencers can be seen as crucial in order for influencer marketing initiatives to achieve the expected results.

The pandemic detonated an increased awareness from consumers on visible individual's behaviors, due to their salient condition. Brands need to take this fact into consideration when establishing formal relationships with them when designing their influencer marketing strategy, which includes elements such as the selection of individuals that could become part of an affiliate program. The purpose of an affiliate program is to track the performance of influencers in terms of turning their audience into customers via links where the brand can validate the source of the recommendation (HubSpot; sprout social).

An example of the unforeseen consequences of incongruent messaging is that of Lea Michelle and her relationship with HelloFresh. Lea Michele is an American actress, who among her latest roles played a character named Rachel Berry in the television series "Glee" (IMDb, n.d.a; IMDb, n.d.b). Michele had a partnership with a meal delivery company called HelloFresh, which was abruptly ended due to statements made by Lea Michelle in regard to the alleged killing of George Floyd at the hands of police, event which was widely publicized, and even detonated a series of protests across various cities in the United States (Taylor, 2020). Remarks made by Michele in regards to the topic detonated responses by other cast members implying that she had been incongruent with the behaviors she was condemning and her own (Wagmeister, 2020) and, shortly thereafter, HelloFresh posted a tweet stating their intolerance towards racism and confirming that they had severed ties with Michele and had no plans to collaborate with her in the future (Morillo, 2020).

Even though this example is unrelated to COVID-19 specifically, it is a clear example of how due to behaviors exhibited by the influencer that are not aligned to corporate guidelines, in this case specifically manifest on values and the own brands' posture with regards to the controversial topic, but exacerbated by the heightened awareness of the audience, resulted in the business relationship to be abruptly terminated, shining even more light into the behaviors deemed unacceptable from the influencer and most likely having a ripple effect on her persona and future brand collaborations.

4.6.1 MANAGERIAL IMPLICATIONS

The COVID-19 pandemic has changed the playing field of influencers. With 78% of consumers expecting brands to help them in their daily lives (Hubspot; Talkwalker), the conversation between brands and their audience will have to evolve, taking into consideration the changes in behavior that the pandemic brought with it, such as KFC dropping their slogan "Finger-Lickin' Good" due to its incompatibility with the behavior that has been recommended to consumers to increase hygiene (BBC News, 2020). Additionally, transactions will have to adapt to a continued distant environment, such as improvement in e-commerce capabilities, which to this day have received a push due to the conditions in order to continue economic transactions, as has been reported by McKinsey, estimating a 3-year acceleration of digitized customer interactions in average (McKinsey and Company, 2020), which could be translated into a continued reliance on digital communication channels to maintain relevance for the intended consumer segment.

Since an influencers' published content is regarded as credible by their audience (Lou and Yuan, 2019) and has been found to have an effect on branded content, brands will still have to be aware of the drawbacks and risks associated with collaborating with influencers, paying special attention to potential spread of misinformation (Ryan, 2020), which has also been a struggle for organizations. Since it's easier to share information than to validate its veracity, it is also easier to share information that could not be accurate. This is where influencers play a determinant role, both by sharing information by reputable sources, and at the same time stopping the continued advance of information that is inaccurate. This would require influencers to assume the responsibility their role has provided them with, as well as on behalf of brands to determine the qualities and profile they want to be associated with (Bailis, n.d.; Harrison, 2018). In the specific case of Arielle Charnas' and Nordstrom, one of the most important facts to consider is that the business relationship they had had already been dissolved by the time Arielle Charnas behaved questionably in terms of her decision making with regards to her Coronavirus diagnosis. This highlights the strength with which the association between two visible parts, in this case Nordstrom and Arielle Charnas, was still in the audiences' minds, even though that relationship had already been finalized. In terms of influencer strategy planning, this should be taken into consideration when making an association visible as well as when an association or relationship is terminated.

Finally, consumers shifted their spending allocation due to the pandemic, in addition to employment cuts in businesses such as hotels, events, dentists, and amusement parks (Pietsch, 2020). Data collected in April 2020 highlighted the categories where consumers were estimating a redistribution of their income, establishing a comparison between 13 countries. According to the report (Jones, 2020), household products like skincare and makeup would see an average reduction of 30% to 50%, as well as restaurants, which would see over 50% reduction in expenditure, while on the other hand groceries, household supplies and home entertainment would experience an increase in consumer expenditure. Brands related to categories where income allocation is reduced need to rethink how to stay relevant in the eyes of their consumers, specifically under the conditions where many are faced with unstable job conditions.

4.6.2 LIMITATIONS

Since the findings are restricted to one case not all the variety of behaviors of an individual that has become salient to an audience can be encompassed. The analysis did not include audiovisual material and focused exclusively on what had been written by the author, which could be considered as a source of bias in terms of the interpretation of such materials, which are the main source of data for the construction of the communication pieces published by news outlets. The reason for not considering these materials and focusing on the articles' author's writing is to highlight the fact that media outlets have an important role in constructing the narrative and interpretation of the facts, and that this also has an impact on constructing the public opinion and interpretation of an influencer's actions, which rely on the diffusion of their content in order to expand their reach beyond their follower base, which in turn is the main point in order to be considered by brands in terms of business associations.

4.6.3 FURTHER RESEARCH

At the time of writing, the COVID-19 pandemic is still ongoing and no definitive treatment has been identified, the public's opinion on the behavior of decision-makers and influencers is still subject to change. Analysis done in retrospect will give each individual a relative weight and influence degree that is most definitely going to determine the whole narrative of the facts. It will be important to learn whether expert organizations such as the WHO played a significant role in instilling specific behaviors congruent with reducing the impact of COVID-19 or not. It will be relevant to learn the criteria the public selected as determinant in providing decision-makers with enough authority in order to follow their advice and recommendations on which behaviors would be most conducive to avoiding contracting COVID-19, as well as the impact of news reports and the constant flow of updates.

4.7 CONCLUSION

The starting point of a successful communication strategy between a brand and their audience is congruence. When congruence is lacking, the audience does not understand the intended purpose of the message, and their behavior will not be what the brand had intended. The issue of congruence when planning and executing an organization's influencer marketing strategy has become the more relevant since its implementation offers benefits that outweigh its challenges, among them believability and less costly compared to other communication channels (von Mettenheim and Wiedman, 2021).

In addition to congruence, relevant factors to take into consideration in terms of influencer strategy design are message value and credibility (Lou and Yuan, 2019). As for how the COVID-19 pandemic has been handled in terms of the information that has been shared and how there are many sources which are constantly expressing opposite views, it comes as no surprise that the outcome of such inconsistent messaging has brought the consequences that we are currently experiencing.

APPENDIX

APPENDIX 4.1 Data Sources

Keyword Set	Headline	Publication Date
Arielle Charnas	Daily News: The latest on something navy, Valentino's next couture show! Plus: Dior's groundbreaking new face!	28.4.2021
	How influencer Arielle Charnas became the face of "Privilege in the Age of the Coronavirus"	2.4.2020
	How influencer Arielle Charnas overcame professional and personal challenges while launching her brand during COVID-19	30.1.2021
	Cosmopolitan Mexico faces backlash over Arielle Charnas cover	2.9.2020

APPENDIX 4.1 *(Continued)*

Keyword Set	Headline	Publication Date
	Fashion influencer Arielle Charnas relaunches something navy line following run at Nordstrom	13.7.2021
	Influencer Arielle Charnas announces sex of baby on the way	7.1.2021
	Arielle Charnas, 30	N/A
	Cosmopolitan Mexico faces criticism after putting influencer Arielle Charnas on cover	3.9.2020
	Is this the end of influencing as we knew it?	3.4.2020
	Influencer Arielle Charnas faces renewed backlash for retreating to Hamptons after COVID-19 diagnosis	3.4.2020
Something navy	Will Arielle Charnas' something navy be a 'Catalyst for Change?'	13.7.2020
	How Arielle Charnas botched the something navy relaunch	17.7.2020
	Arielle Charnas Opens first something navy boutique in New York — Here's everything you need to know	31.8.2020
	It might just sellout (again)—influencer Arielle Charnas just launched her debut something navy namesake collection	13.7.2020
	Something navy starts its second life as a DTC brand	13.7.2020
	An influencer just sold more than $1 million in clothing in 30 minutes, even after recent backlash over fleeing NYC following a COVID-19 diagnosis	13.7.2020
	Influencer Arielle Charnas accused of copying design for something navy	15.5.2020
	Arielle Charnas's something navy collection is here and we've already picked our faves.	13.7.2020
Arielle Charnas COVID drama	The biggest influencer controversies of 2020	24.12.2020
	A timeline of something navy influencer Arielle Charnas's COVID-19 D-r-a-m-a	2.4.2020
	Did Arielle Charnas buy a bunch of new followers this week?	9.4.2020

APPENDIX 4.1 *(Continued)*

Keyword Set	Headline	Publication Date
	The saga of Arielle Charnas, patient zero of influencer-coronavirus controversy	30.4.2020
	Arielle Charnas' Husband Brandon apologizes for 'offensive' comments after intense backlash	8.5.2020
	Arielle Charnas denies faking coronavirus results as she is criticized for fleeing to Hamptons	3.4.2020
	Arielle Charnas's Coronavirus saga is a rabbit hole impossible not to fall into	1.4.2020
	'Covidiot' blogger Arielle Charnas may have ruined her brand	2.4.2020
	One year later, and Arielle Charnas still doesn't get it	22.3.2021

Source: Elaborated by author based on Google Search results.

KEYWORDS

- **business associations**
- **congruence**
- **COVID-19 pandemic**
- **data analysis**
- **managerial implications**
- **Nordstrom disassociation**

REFERENCES

Bailis, R., (n.d.). *The State of Influencer Marketing: 10 Influencer Marketing Statistics to Inform Where You Invest*. Retrieved from: BigCommerce: https://www.bigcommerce.com/blog/influencer-marketing-statistics/#what-is-influencer-marketing (accessed on 18 February 2022).

BBC News, (2020). BBC News. *KFC Drops Finger-Lickin' Good Slogan amid Coronavirus*. Retrieved from: https://www.bbc.com/news/business-53901236#:~:text=Global%20

fast%20food%20giant%20KFC,because%20of%20the%20coronavirus%20pandemic (accessed on 18 February 2022).

Bengtsson, M., (2016). How to plan and perform a qualitative study using content analysis. *NursingPlus Open, 2*, 8–14.

Bowman, S. N. (2020). *Influencer Versus Body Positive Marketing: Perceived Brand Personalities in the Eye of the Target Consumer* [Doctoral dissertation]. https://trace.tennessee.edu/utk_chanhonoproj/2335 (accessed on 15 March 2022).

Caramanica, J., (2020). New York Times. *This 'Imagine' Cover is No Heaven.* Retrieved from: https://www.nytimes.com/2020/03/20/arts/music/coronavirus-gal-gadot-imagine.html (accessed on 18 February 2022).

Centers for Disease Control and Prevention, (2020a). CDC 24/7: Saving lives, protecting people. *Human Coronavirus Types.* Retrieved from: https://www.cdc.gov/coronavirus/types.html (accessed on 18 February 2022).

Centers for Disease Control and Prevention, (2020b). CDC 24/7: Saving lives, protecting people. *When to Quarantine: Stay Home if You Might Have Been Exposed to COVID-19.* Retrieved from: https://www.cdc.gov/coronavirus/2019-ncov/if-you-are-sick/quarantine.html (accessed on 18 February 2022).

Chopra, A., Avhad, V., & Jaju. S., (2020). Influencer marketing: An exploratory study to identify antecedents of consumer behavior of millennial. *Business Perspectives and Research,* 1–15.

Crain's New York Business, (n.d.). Crain's New York Business. *40 Under 40 Class OF 2018.* Retrieved from: https://www.crainsnewyork.com/awards/40-under-40-class-2018 (accessed on 18 February 2022).

Delisle, M. P., & Parmentier. M. A., (2016). Navigating person-branding in the fashion blogosphere. *Journal of Global Fashion Marketing, 7*(3), 211–224.

DeSimone, M. T., (2018). Insta-famous: Challenges and obstacles facing bloggers and social media personalities in defamation cases. *The Modern American, 11*(1), 70–101.

Emmanuele, J., (2020). Bustle. *Gal Gadot Understands Why People Didn't Like Her "Imagine" Video.* Retrieved from: https://www.bustle.com/entertainment/gal-gadot-responded-to-the-imagine-video-backlash (accessed on 18 February 2022).

Finne, A., & Grönroos. C., (2017). Communication-in-use: Customer-integrated marketing communication. *European Journal of Marketing, 51*(3), 445–463.

Gaur, A., & Kumar. M., (2018). A systematic approach to conducting review studies: An assessment of content analysis in 25 years of IB research. *Journal of World Business, 53*, 280–289.

Google Inc., (2021). *Google Trends.* Retrieved from: https://trends.google.com/trends/?geo=MX (accessed on 18 February 2022).

Grand View Research, Inc., (2020). PR Newswire. *Influencer Marketing Platform Market Size Worth $23.52 Billion by 2025: Grand View Research, Inc.* Retrieved from: https://www.prnewswire.com/news-releases/influencer-marketing-platform-market-size-worth-23-52-billion-by-2025-grand-view-research-inc-300985412.html (accessed on 18 February 2022).

Haradhan, M., (2018). Qualitative research methodology in social sciences and related subjects. *Journal of Economic Development, Environment and People, 7*(1), 23–48.

Harrison, K., (2018). Forbes. *Want to Try Influencer Marketing? Be Careful.* Retrieved from: https://www.forbes.com/sites/kateharrison/2018/05/22/

want-to-try-influencer-marketing-be-careful/?sh=184c4e884a8b (accessed on 18 February 2022).

Heale, R., & Twycross. A., (2018). What is a case study? *Evidence-Based Nursing, 21*, 7–8.

Hubspot, (2020). *Not Another State of Marketing Report: Trends and Data from Over 3,400 Global Marketers Plus Progressive Strategies from Our Experts and Partners.* Hubspot.

HubSpot; Sproutsocial. (n.d.). The Ultimate Guide to Influencer Marketing. (2020). Hubspot, Inc. https://offers.hubspot.com/influencer-marketing-guide (accessed on 15 March 2022).

Hubspot; Talkwalker. (n.d.). *Tendencias de Las Redes Sociales 2021: Un Informe Detallado Para Impulsar tu Estrategia de Redes Sociales.* (2021). Hubspot, Inc. https://f.hubspotusercontent00.net/hubfs/53/00-OFFERS-HIDDEN/%5BSPANISH%5D%20SMT%202021/Tendencias%20Redes%20Sociales%202021.pdf?submissionGuid=715f6e43-ddd8-41ff-8803-224acc8b671d (accessed on 15 March 2022).

IMDb, (n.d.). *Glee*. Retrieved from: https://www.imdb.com/title/tt1327801/?ref_=nm_flmg_act_6 (accessed on 18 February 2022).

IMDb, (n.d.). *Lea Michele*. Retrieved from: https://www.imdb.com/name/nm0584951/?ref_=nv_sr_srsg_0 (accessed on 18 February 2022).

IMDb.com. (n.d.). Internet movie database. *Gal Gadot*. Retrieved from: https://www.imdb.com/name/nm2933757/?ref_=nv_sr_srsg_0 (accessed on 18 February 2022).

Influencer Marketing Hub, (2021). Retrieved from: https://app.influencermarketinghub.com/tools/instagram/fake-followers-checker/ (accessed on 18 February 2022).

Jin, S. V., Muqaddam, A., & Ryu. E., (2019). Instafamous and social media influencer marketing. *Marketing Intelligence & Planning.*

Jones, K., (2020). Visual capitalist. *How COVID-19 Consumer Spending is Impacting Industries*. Retrieved from: https://www.visualcapitalist.com/consumer-spending-impacting-industries/ (accessed on 18 February 2022).

Keller, K. L., (2016). Unlocking the power of integrated marketing communications: How integrated is your IMC program? *Journal of Advertising, 45*(3), 286–301.

Ki, C. W., Cuevas, L. M., Chong, S. M., & Lim. H., (2020). Influencer marketing: Social media influencers as human brands attaching to followers and yielding positive marketing results by fulfilling needs. *Journal of Retailing and Consumer Services, 55.*

Laurie, S., & Mortimer. K., (2019). How to achieve true integration: The impact of integrated marketing communication on the client/agency relationship. *Journal of Marketing Management, 35*(3, 4), 231–252.

Lee, A., (2020). *Ellen DeGeneres Sparks Backlash After Joking that Self-Quarantine is Like 'Being in Jail'.* (CNN) CNN entertainment. Retrieved from: https://edition.cnn.com/2020/04/08/entertainment/ellen-degeneres-quarantine-jail-trnd/index.html (accessed on 18 February 2022).

Lee, K., (2018). Marketing land. *2018: Year of the Influencer or Year of the Influencer Marketing Implosion?* Retrieved from: https://marketingland.com/2018-year-influencer-year-influencer-marketing-implosion-231617 (accessed on 18 February 2022).

Lou, C., & Yuan. S., (2019). Influencer marketing: How message value and credibility affect consumer trust of branded content on social media. *Journal of Interactive Advertising, 19*(1), 58–73.

Magno Hack S. R. L. (n.d.). *Analyze any Instagram profile! Analyze any Instagram profile!*. Retrieved July 12, 2021, from https://business.notjustanalytics.com/plus/ariellecharnas (accessed on 18 February 2022).

Martinez-Lopez, F. J., Anaya-Sanchez, R., Fernandez, G. M., & Lopez-Lopez. D., (2020). Behind influencer marketing: Key marketing decisions and their effects on followers' responses. *Journal of Marketing Management, 36*(7, 8), 579–607.

McKinsey & Company, (2020). McKinsey & Company. *Digital Fast-Forward: Homebound Consumers Forced Companies to Jump Years Ahead.* Retrieved from: https://www.mckinsey.com/featured-insights/coronavirus-leading-through-the-crisis/charting-the-path-to-the-next-normal/digital-fast-forward-homebound-consumers-forced-companies-to-jump-years-ahead (accessed on 18 February 2022).

Moore, A., Yang, K., & Kim. H. M., (2018). Influencer marketing: Influentials' authenticity, likeability and authority in social media. *ITAA Proceedings* (Vol. 75, pp. 1–3). Cleveland: International Textile and Apparel Association, Inc.

Morillo, A., (2020). delish. *HelloFresh Has Ended Its Partnership with Lea Michele in Response to Her Alleged Racially Charged Behavior.* Retrieved from: https://www.delish.com/food-news/a32755714/hellofresh-ends-lea-michele-partnership/ (accessed on 18 February 2022).

Morteo, I., (2017). Influencers: An account of their role as strategic spokesperson.

Pietsch, B., (2020). Business insider. *20.5 Million People Lost Their Jobs in April. Here Are the 10 Job Types that Were Hardest Hit.* Retrieved from: https://www.businessinsider.com/jobs-industries-careers-hit-hardest-by-coronavirus-unemployment-data-2020-5?r=MX&IR=T (accessed on 18 February 2022).

Ryan, J., (2020). c|net. *Coronavirus Panic Thrives on Twitter, and Science Struggles to Keep Up.* Retrieved from: https://www.cnet.com/news/coronavirus-panic-thrives-on-twitter-and-science-struggles-to-keep-up/ (accessed on 18 February 2022).

Stankevich, A., (2017). Explaining the consumer decision-making process: Critical literature review. *Journal of International Business Research and Marketing, 2*(6), 7–14.

Taylor, D. B., (2020). The New York Times. *George Floyd Protests: A Timeline.* Retrieved from: https://www.nytimes.com/article/george-floyd-protests-timeline.html (accessed on 18 February 2022).

The Atlantic Monthly Group, (2020). The COVID tracking project. *The Public Deserves the Most Complete Data Available about COVID-19 in the US. No Official Source is Providing it, so we Are.* Retrieved from: https://covidtracking.com/ (accessed on 18 February 2022).

Von, M. W., & Wiedman. K. P., (2021). The complex triad of congruence issues in influencer marketing. *Journal of Consumer Behavior,* 1–20.

Wagmeister, E., (2020). Variety. *Lea Michele Dropped by HelloFresh After 'Glee' Co-Star Samantha Ware's Accusations.* Retrieved from: https://variety.com/2020/tv/news/lea-michele-hellofresh-dropped-samantha-ware-accusations-1234623622/ (accessed on 18 February 2022).

Weismueller, J., Harrigan, P., Wang, S., & Soutar. G. N., (2020). Influencer endorsements: How advertising disclosure and source credibility affect consumer purchase intention on social media. *Australasian Marketing Journal,* 1–11.

World Health Organization, (2020a). World Health Organization. *Timeline of WHO's response to COVID-19*. Retrieved from: https://www.who.int/news/item/29-06-2020-covidtimeline (accessed on 18 February 2022).
World Health Organization, (2020b). World Health Organization. *Timeline: WHO's COVID-19 Response*. Retrieved from: https://www.who.int/emergencies/diseases/novel-coronavirus-2019/interactive-timeline#! (accessed on 18 February 2022).
Zhu, H., Xie, H., & Veeck. A., (2020). Top macromarketing lessons of the COVID-19 outbreak. In Barrios, A., Fowler, G., Shultz, C., & Gentry, J., (eds.), *Transitioning Markets: Opportunities, Challenges & Future Trends* (pp. 39–41). Bogotá: Macromarketing Society, Inc.

CHAPTER 5

"It's Not Panic, I'm Just Getting Prepared": Exploring Changes in Consumer Behavior During the COVID-19 Pandemic

TERESA TREVIÑO

Business School, University of Monterey, San Pedro Garza García, Mexico, E-mail: teresa.trevinob@udem.edu

ABSTRACT

Novel coronavirus disease 2019 (COVID-19) has changed people's lifestyles, activities, and priorities overnight. New needs have emerged, resulting in changing demands for products and services. From a business perspective, the coronavirus pandemic (COVID-19) has forced companies to adapt and innovate, finding new ways in which their messages and products can reach customers. However, to successfully do so, the first step is to understand the consumer in the light of the pandemic. Therefore, this chapter contributes to the literature by shedding light on how consumers have modified their retail shopping behavior during the COVID-19 pandemic. Specifically, the research addresses a) consumer experiences with the pandemic and confinement; b) changes in consumers' shopping activities; and c) consumer perceptions on how companies and brands have managed the situation to support consumers. Overall, the results of this research suggest that consumers are more informed, prepared, critical, and connected, and therefore prefer to relate to brands that seem caring, emphatic, and honest. Managers can benefit from these results by

Marketing by Contingency in the Time of COVID-19: Overcoming Business Crises and Meeting Marketing Challenges. Alicia De La Peña, PhD & Juan Bernardo Amezcua Nunez, PhD (Eds.)

understanding the motivations of consumers to buy, their perceptions of brand communications during this time, and, more importantly, consumer expectations in such complex times.

5.1 INTRODUCTION

During the initial months of 2020, an announcement was unexpectedly made in many countries around the world: novel coronavirus disease 2019 (COVID-19) will become a global pandemic. Since then, we have been facing this unprecedented event that has drastically impacted our lives.

Our lifestyles, activities, and priorities have changed abruptly, and without notice, things that once were taken for granted—like hugging or kissing someone you love—are now more valued and missed. There is no question that people around the world are trying to adapt to the new circumstances. In many countries, confinement has been mandatory at some point, to prevent the spread of the disease. Whenever possible, work, and school activities have been shifted to home and online context to maintain social distance. As a result, we have witnessed unprecedented sights, such as important global destinations that were once busy and crowded 24/7 become empty and desolate. For example, there has probably never been a time when we could witness empty streets at Japan's Shibuya or the Las Vegas strip or see clear waters with dolphins in Italy's Venice canals. As the months pass by, and people wait for a solution, the reality is that people were forced to change overnight. New needs emerged from the situation, resulting in changing demands for products and services. Furthermore, it is known that when facing a crisis, people's behavior, specifically their purchasing habits, change in a way that enables them to gain some sense of control over the situation (Mathur, Moschis, and Lee, 2003; Zwanka and Buff, 2020).

From a business point of view, the coronavirus pandemic (COVID-19) has required companies to adapt and innovate, searching for new ways for their messages and products to reach customers. The damaged economy, increased competition due to the expanded use of online platforms, and changes in consumer behavior have made these months difficult to bear. Given that a pandemic of this scale has not been seen in recent times, so there is little research addressing the changes in consumer behavior that a health crisis of this magnitude might precipitate.

However, research from past stressful events has demonstrated that the effect on people due to crises occurs both during the event and years after it ends (Zwanka and Buff, 2020). This suggests that the impact of an uncertain, difficult, or stressful event might cause long-term effects on consumer behavior.

Therefore, this chapter contributes to the literature by shedding light on how consumers modified their retail shopping behavior during the COVID-19 pandemic. Specifically, the participants discussed (a) their experiences with the pandemic and confinement since March 2020, (b) the changes they had to make in their shopping activities to adapt to their new circumstances while keeping themselves and their families safe, and (c) their perceptions of how companies and brands managed the situation to support consumers.

It can be expected that even when the coronavirus disease is under control and life returns to "normal," that some of these changes in consumer behavior will persist (Mathur, Moschis, and Lee, 2003; Zwanka and Buff, 2020). Some scientists have predicted that the chances of new global pandemics are increasing (Global Preparedness Monitoring Board, 2019; Kirk and Rifkin, 2020). Therefore, future research on this topic can contribute to understanding the changes in consumers' lives and how businesses can adapt to a crisis that is likely to prevail.

This chapter is structured as follows. First, a literature review addresses previous research on how consumers and companies have changed during past crises. Then, the methodology as well as the data collection and analysis process adopted by this research is explained. Next, the results are presented and discussed. Finally, the chapter concludes by exploring managerial implications, limitations, and future research lines.

5.2 LITERATURE REVIEW

5.2.1 CONSUMER BEHAVIOR DURING A CRISIS

Research concerning consumer behavior during difficult or uncertain times has been addressed before in the literature. Literature has established that when consumers experience life events that create stress and require new roles and dynamics, they modify their consumption lifestyles and brand preferences in an attempt to adapt to new circumstances (Mathur,

Moschis, and Lee, 2003). Examples of such large-scale life events include earthquakes, hurricanes (e.g., Hurricane Katrina in 2005), terrorist attacks (e.g., 9/11), complex political times, health concerns, such as epidemics, among others (Zwanka and Buff, 2020). As expected, stressful events and difficult situations cause consumers to change their consumption habits as a way to handle stress and increase their perceived control (Mathur, Moschis, and Lee, 2003; Zwanka and Buff, 2020).

Additionally, the literature has suggested that changes in product acquisition and panic purchasing are two activities that provide peace of mind to some consumers (Ballantine, 2013; Yuen, Wang, Ma, and Li, 2020). Panic buying is characterized by consumers buying unusual or excessive amounts of products. These decisions might occur when consumers anticipate specific events such as disasters, future price increases, and limited availability, among others (Loxton et al., 2020).

Research has suggested that when consumers believe that there is a high risk of contracting a disease, their actions are motivated by the desire to protect themselves and their families. Following a recent meta-analysis conducted by Yuen et al. (2020), four causes of panic buying have been investigated in the literature. The first cause is the perception of threat and scarcity. During such times, possessing certain products helps to reduce the perceived risk of not having them or not being able to buy them at a later time. Therefore, it is expected that an increase in demand in certain categories of products will be observed. Second, this behavior is often related to a lack of information and fear of the unknown (Yuen et al., 2020). Consumers translate this fear into modified purchasing patterns that provide them with a sense of comfort and security (Loxton et al., 2020). Third, this meta-analysis suggested that panic buying might function as a coping mechanism. The acquisition of products in an unprecedented manner might compensate for the psychological losses that consumers experience during uncertainty (Yuen et al., 2020). Finally, social influence is another factor that shapes consumer behavioral changes during difficult times. As social beings, humans are naturally driven to act following others and, in this context, their purchases might be affected by the opinions, actions, and perceptions of society. When some behaviors are undertaken by a large group of people or displayed on social or traditional media, consumers might be moved to engage in such activities as well. Particularly for the COVID-19 pandemic, social media has played an important role in disseminating information that has resulted in collective anxiety and panic.

When rumors of a possible shortage spread throughout Facebook, Twitter, or other popular social networking sites, people immediately begin buying in large quantities, and eventually, the potential shortage becomes a reality (Freeland, 2020).

When individuals align their actions and thoughts to those of a social group, this behavior has been described as a "herd mentality" in the literature (Loxton et al., 2020). During a crisis, this phenomenon is manifested in several ways such as panic buying and collective social anxiety, and it can even be reflected in stock prices. One outcome of such collective behavior is that it can unexpectedly modify the market, as the demand for some products suddenly increases, often to be stored unnecessarily in a consumer's home. Since supply cannot adjust quickly enough, other problems arise, such as changes in product prices due to scarcity, store policies limiting purchases per customer, individuals reselling products at unprecedented prices, all of which ultimately generates more panic purchasing.

We have learned that consumers have changed in many ways and that brands have been forced to change accordingly to adapt to these new circumstances. As consumers are still facing this unprecedented pandemic, things have been put into perspective, and many individuals are rethinking their actions and assessing what is important for themselves. This includes consumers' brand relationships.

The present research attempts to continue the conversation on the topic and attempt to identify how consumers have modified their retail shopping behavior during the COVID-19 pandemic.

5.2.2 BRAND PERCEPTIONS AND CONSUMER CHANGE

A recent survey conducted with individuals around the world demonstrated that consumers and their future purchase decisions are influenced by how brands responded to the pandemic. Some consumers have already reported a change in their brand preferences based on their perception of brands' reaction and response to the crisis (Kirk and Rifkin, 2020).

Companies around the world are facing tough economic times as the resulting lockdown negatively impacted many businesses. Brands that have been successfully surviving these challenging times were forced to quickly adapt their marketing strategies, often resorting to innovative means for delivering products and services to consumers and relying on the Internet to reach consumers confined at home. Social media has played an important role in consumer communication. However, brands can be vulnerable to negative perceptions based on their crisis response. Recent research suggests that consumers care about how brands respond to the crisis and brands that acted selfishly or failed to make sacrifices to help others cope with this situation can create negative consumer perceptions (Kirk and Rifkin, 2020). Furthermore, consumers expect brands to show awareness and compassion on the impact of the COVID-19 pandemic and find ways for their products to help people cope (Rogers, 2020). There are many examples of brands responding favorably to the COVID-19 crisis. Global brands such as Coca-Cola, Audi, and McDonald's were the first to modify some part of their logo—usually by physically separating elements in their logos—to promote the social distancing recommended to stop COVID-19's spread. Other examples of company actions include Google provided their Google Hangouts Meet service free to members to contribute to maintaining open communication between people during confinement, Amazon donated approximately 8,200 laptops to help students without devices continue online schooling, and Microsoft maintained its hourly-paid workers' pre-pandemic income despite having to work fewer hours (Littleton, 2020). In Mexico, brands have displayed positive reactions to the situation. Grupo Modelo donated 300,000 sanitizing gel bottles produced with the alcohol extracted when manufacturing beer. When closing their movie theaters for the pandemic, Cinepolis—the biggest cinema company in the country—displayed on the billboards where movies titles are usually advertised, exhibited the message, "Movies have shown us that there is always a happy ending, we are going to miss you. Take care" (Forbes, 2020).

Past research on consumer relationships with brands suggests that consumers create psychological contracts with the brands they like (Montgomery, et al., 2018). In this contract, the terms of the relationship are written implicitly, and when brands behave differently than expected—i.e., violate the contract—there can be negative consequences for the relationship. This outcome is especially true for committed consumers engaged in a strong brand relationship (Montgomery et al., 2018). We have also discovered that brands are created by offering much more than just the functional benefits of their products and services; their values and how they connect with consumers are also relevant. Therefore, brands not exhibiting compassion and taking an interest in their employees and society as a whole during this pandemic crisis could fracture the psychological contract with consumers.

Coping with this unprecedented moment will certainly be different for every brand. However, thus far the conversation has moved towards the proposition that brands that respond with a meaningful approach have created positive perceptions by giving consumers a reason to believe that we can overcome this crisis together. This research attempts to dig deeper into consumer's perceptions of brand communications during this time of crisis as well as understand their expectations in terms of the brand's actions.

5.3 METHODOLOGY

The present research adopts a qualitative, exploratory approach to shed light on changes in consumer behavior resulting from this pandemic. Considering that the objective of the research is to understand changes in consumer behavior, qualitative methods allow to obtain insights from participants, as well as to understand the underlying motives of their actions. The data collection process consisted of two phases. The first phase included in-depth interviews since this methodology is flexible enough to address personal topics with participants regarding their reflections on how the pandemic has reshaped their lives. This method was possible to conduct even in the difficult situation of confinement. A total of 22 in-depth interviews were conducted with participants, including 12 men and 10 women, between 17 and 57 years of age, 32 on average. Considering the current social distancing recommendations due to the COVID-19 pandemic, all interviews were held online using Zoom and recorded both the audio and video with the participant's acceptance and lasting an average of 30 minutes. All interviews were conducted in Spanish, and therefore, the provided quotations in this chapter are the translation of participant's comments. All participants belong to medium to medium-high socioeconomic level. There was diversity in the occupations of the participants; we interviewed students, employees, and housewives (see Table 5.1).

The second phase of the data collection process included distributing a survey online that included open-ended questions to obtain additional information about people's purchasing patterns during the pandemic. It focused specifically on analyzing the products that were purchased more frequently, products that were not consumed, and reasons why people engaged in panic purchasing. This survey was answered by a convenience sample of 75 participants, with an average age of 37, of which 66 were women and nine were men. All participants, for both the interviews and survey, belonged to a medium to medium-high socioeconomic status and lived in a city in northern Mexico.

TABLE 5.1 Participant Summary

Interview	Participant Name	Age	Occupation
1	Abraham T.	51	Employee
2	Briana G.	20	Student
3	Carlos E.	20	Student
4	Ana B.	26	Employee
5	Claudia S.	51	Employee
6	Cristina L.	44	Housekeeper
7	Edgar L.	47	Employee
8	Fernando L.	34	Employee
9	Fernando V.	57	Employee
10	Joaquin B.	17	Student
11	Laura S.	19	Student
12	Leonardo M.	21	Employee
13	Manuel V.	22	Student
14	Marcelo P.	21	Student
15	Mariana D.	20	Student
16	Mariana M.	50	Housekeeper
17	Mauricio D.	20	Student
18	Pablo C.	23	Student
19	Pablo J.	20	Student
20	Rocío B.	46	Employee
21	Sofía G.	22	Student
22	Viviana C.	57	Housekeeper

Source: Own elaboration.

During the data analysis process, the transcripts of the interviews were created and examined for emergent categories. The coding and categorization phase was an iterative process that enabled the researcher to analyze the participant's narratives in a deeper form. Later, the codes between the participants were crossed to find patterns and relationships among them. Then, the textual answers from the online survey were downloaded and coded into product categories and motives to increase purchases during the pandemic. Finally, both datasets were analyzed together to identify similarities and differences and conclude with more generalizable findings. The use of two different data collection methods helped to triangulate the results and ensured greater validity.

5.4 RESULTS

5.4.1 LIFE DURING THE LOCKDOWN

Participants reflected on how life drastically changed as a result of the COVID-19 pandemic and the required lockdown in response to it. In Mexico, most activities were halted around March 15, 2020, so participants have been reducing their outdoor activities ever since. On one hand, younger participants that once had an established daily routine for their school activities explained how their day is now spent in front of the computer. All classes and lectures that were once taken in the classroom are now conducted online. Even extracurricular activities such as art, music, dance, and sports are now also held online. This has been a difficult situation, as this group of participants was accustomed to being outside their homes almost all day long, as one participant reflected:

"The first two weeks... were like a vacation... but then, it has been really difficult. I think that this [taking online classes] really affects me a bit, emotionally speaking, because it sometimes is really exhausting to be all day here on the computer, sitting all day" (Carlos, 20).

The lack of contact with friends at this age is especially difficult, as younger students previously went out to restaurants, bars, and parties on a regular basis and suddenly one day it was all over:

"It has been difficult, to tell you the truth. I am very used to being busy, I was never at home, I was just here to eat and sleep, but really not much here. But the fact that now I have to be here and not gone out anywhere, life is really different from what I was used to, and not seeing my friends also impacts a lot" (Briana, 20).

"It has been difficult… it really has been difficult; I have spoken with [several] psychologists through video calls because it has not been easy… This has been a lot of confinement… and being with yourself. Not letting your mind play you" (Leonardo, 21).

On the other hand, older participants explained how their activities had to change due to this situation. Considering that most of them have to work, they might still have to be out of their homes during the day. However, most participants agreed that most of their work activities had migrated to the online context and that interaction with other people had decreased significantly. These participants sometimes felt confused and stressed because the line between work time and personal time had become blurred.

"It has been such a strong change. Really, to learn how to be organized with my time. To be able to divide work time and family time, because we are interacting a lot in our home, but we need to realize that is the time of our company, and I have to be very strict with this time" (Fernando, 57).

Additionally, not all changes are negative, and some participants focused on the positive aspects of the lockdown. Being at home with no place to go has given some participants the time to focus on things that their usual daily activities caused them to forget or leave behind. One activity that was mentioned repeatedly was attending to household tasks such as small repairs, organizing things, and taking care of details around the house. Furthermore, some older participants reflected on the positive side of being at home with their children, as this confinement at home has provided an opportunity to bond more.

"In this time where time to spare, we have been doing things that you leave with time. Things that you think 'I'll do later.' We have been arranging pending things here at home. Also, this time has allowed me to interact with my children. I knew what they wanted obviously, but at this time, we have had long talks" (Claudia, 51).

Another issue that emerged in this conversation was the Internet available for all family members being connected at the same time. Some participants reflected on how this has been a topic of discussion among friends and colleagues, considering that all family members have online activity during the day, requiring bandwidth to make video calls and attend classes and meetings.

"Some of my friends [have told me] they had new rules in their homes. Forget about YouTube, the person that has to be in class is the priority, the person that needs to be in meetings or calls is also a priority, all the others, need to wait [to use the Internet]. It has been a conflict" (Edgar, 47).

Overall, and despite their ages and occupations, all participants agreed that the lockdown has changed their lives drastically, without notice. It has been a process of ups and downs, realizing that this is not something that will be over in the next couple of days, and making adjustments as the days—and months—are passing. From a marketing perspective, all of these changes in their lives serve as a basis for understanding the changes they are facing as consumers. As their lifestyles drastically changed, and time is destinated to other activities, participants recognized that new needs in products and services emerged. This will be addressed in the next section.

5.4.2 CHANGES IN PURCHASING PATTERNS

On one hand, participants reflected on the new products that they have bought during these months of confinement. Most products are related to new needs that are present due to the current pandemic situation. These products can be classified as things to protect, prevent, or treat COVID-19.

Considering that most people have more time to be home, new products also include things that make housework easy and enjoyable—like automatic vacuums cleaners and pre-cooked meals—and products that help people entertain, stay healthy, and pass the time, such as exercise equipment, books, and anti-stress products. Furthermore, they reported the need to buy products related to equipping home offices and conducting homeschooling. This includes computers, laptops, tablets, desks, and chairs, earphones, and computer supports. Moreover, participants reported products related to home improvement, like plants for the garden, paint, among others. As one participant mentioned:

"One weird purchase I recently made is dirt and grass to plant on my patio. I wouldn't be able to do that if it wasn't for COVID. It was something I had in mind but it wasn't my priority... now I have time to do it. If it was a normal day, I would go with my friends to a bar" (Fernando, 34).

On the other hand, there are also products and services that participants stop consuming during these months. Most of them are related to things that are consumed outside of the home, like going out to restaurants, bars, cinemas, and casinos. Other goods include hair, makeup, and nail services; clothes and shoes; airplane tickets; and gasoline.

5.4.2.1 PANIC PURCHASING

It was very interesting to address the topic of panic buying with participants. When asked directly if they had made one or more panic purchases during the pandemic, all of them answered that they had not. However, as the discussion continued, the question was later rephrased, and the answers obtained gave a clear indication of panic buying. For example, most participants mentioned that they had increased their consumption of certain products during the pandemic, for several reasons. To be prepared for the future, for fear that the product would not be available in the future, or because prices might increase in the future were some of the most common reasons provided. For example, the following quotes of two of the participants reflected this phenomenon:

TABLE 5.2 Products with Increased Consumption During the COVID-19 Pandemic

Product	Examples
Sanitizers	Wipes, sanitizers, antibacterial gel, and spray
Cleaning products	Bleach, floor cleaners, soaps
Food	Canned food, fruits, and vegetables, meat to freeze
COVID-19 protection	Facemasks, face shields, gloves
Personal hygiene	Hair and face products
Toilet paper	–
Bottled water	–
Clothes	Pajamas, exercise clothing, socks
Vitamins	Multivitamins, vitamin C, probiotics
Books	–
Snacks and desserts	Chips, bread, cakes
Technology	Computer accessories, cell phones
Cigarettes	–

Source: Own elaboration.

"No, I haven't made panic purchases. When I went to the supermarket, I did buy more canned food, tuna, mashed potatoes in boxes, tomato sauce in boxes, things that normally I don't buy… but it wasn't a panic purchase… I did it to have it just in case it runs out [on the stores], to be prepared… (Cristina, 44).

"At the beginning [of the pandemic] I bought… but I didn't buy in excess, but it was like, let's buy a little bit more of everything just in case it runs out" (Edgar, 47).

There are many similar narratives by other participants conveying the idea that when faced with the question of panic purchasing, they prefer not to classify their behavior with that term. However, the behavior that they described represents a theoretical description of panic buying. The products that were reported to have been purchased in increased quantities are listed in Table 5.2.

When analyzing the results of the online survey, the reasons underlying the behavior of panic buying can be found in Table 5.3. As expected, anxiety, perceived scarcity, and fear are the top three motivational factors that prompted an increase in the quantity of some products normally purchased by consumers.

TABLE 5.3 Reasons Prompting Panic Buying by Consumers

Motives for Panic Buying	
Anxiety	Future lack of money
Perceived scarcity	To see what people do
Fear	Stress
Free time	Depression
Uncertainty	Lack of information
To be prepared/for security	

Source: Own elaboration.

Lastly, consumers reported more frequent online purchasing behavior, motivated by the fact that at this moment, it is difficult to physically visit stores. The most mentioned online store was Amazon.com, followed by Mercadolibre.com and clothing stores. They reflected on the fact that it is important to have a friendly website or app that makes the online purchasing process quick and easy. Social networking sites were also mentioned as a place where consumers got to know new products and services during this moment of confinement.

5.4.3 CONSUMPTION OF COVID-19 CONTENT AND NEWS

During the interview, participants reflected on the online content to which they are exposed every day. As more time is spent on the computer, there is also more time to consume online information through social networking sites and websites. Participants mentioned how COVID-19 news and posts are abundant on their social networking sites. In general, they discussed how the information regarding the COVID-19 pandemic is mostly negative news. Death rates, people suffering, and things that are occurring in the country or around the world were some of the examples offered.

The interviewed participants also exhibited an aversion to such news, considering that it has been several months since the beginning of the pandemic. For example, the following participant reflected on this issue:

"I prefer other types of content, I see [COVID news] to be informed, but only for a little while and then you get depressed... you say: 'enough.' Then you search for other types of content... I search for [other places/sites] that have more happy news. I read a little bit [about COVID], I find out how the situation is, and then I read about other things" (Cristina, 41).

Additionally, it was mentioned that consuming such a large amount of negative news every day for many months had negatively impacted their overall well-being and mental health. Therefore, they had realized this fact and had taken some action to prevent feeling overwhelmed:

"I need to be informed, but to one extend I think this information is also negative to yourself. So, I try to distract myself and look for other types of information. I want to know the basics, the phase we are in, the number of cases, etc., but it is also very possible to find fake news. We need to be informed but not overwhelm us and create general panic" (Mauricio, 20).

"I try to avoid it, really [COVID-19 related news] because it overwhelms you. To see that all day is bad... it makes bad to you... as you think all day about that is depressing" (Leonardo, 21).

The conversations with participants moved around how at first, they had an optimistic attitude regarding the end of the pandemic and resuming their normal activity in the short term. However, as time went on, this positive and hopeful attitude has been declining. In general, people have been facing this situation for approximately nine months, so most of the participants are certain that this situation will not end in 2020. As one participant mentioned:

"At first, I had positive thoughts, that this will not last that long, that it will end quickly…[ ]… but each time… is taking longer [the pandemic]. I don't like to see news about how we are doing [as a country] it's something that gets me even more frustrated" (Briana, 20).

To cope with this saturation of COVID-19-related news, participants are taking some actions to consume other types of online content that might help them during this difficult time. It was especially common to hear that they were taking online classes, workshops, or conferences about personal—not work-related—topics that they enjoyed. For example, the most often mentioned topics were religion, personal growth, cooking classes, exercise, games, movies, and documentaries.

"[I see content about] exercise, and easy cooking—because you are all day at home… or games that can, in a way, entertain you… there are many topics that you can find every day. So that type of content is present in my day" (Laura, 19).

5.4.4 CONSUMER PERCEPTIONS OF BRANDS DURING THE EMERGENCY

Continuing with the online content topic, discussions with participants then moved towards their perceptions on how brands have addressed the pandemic and the content that they have shared during these months. Most interviewees were aware that some companies had changed their communications and had touched on the COVID-19 topic at some point.

Most of the mentioned communications were around campaigns to help inform the community about the virus, its symptoms, how it can be spread, and the basics about how to prevent transmission.

Similarly, other mentioned examples concern how companies assist their employees during this difficult time. For example, modifying working hours, providing protective equipment to prevent virus transmission, continuing to pay their salaries even though the business is closed, among others.

"I saw some campaigns [of companies] in which they are explaining how they help their employees… so that they can work less [time] so that there are not so many people in the offices. They also give them support and try to avoid letting people go. They are also giving them help in the transportation topic so that employees can have less risk when going [from and to] work" (Abraham, 51).

Furthermore, some younger participants—ones that use social media heavily—mentioned how companies are using influencers, celebrities, and athletes to increase awareness of their support campaigns around the COVID-19 topic, such as their campaigns about staying at home to maintain social distance and about funds and donations of food and medical supplies to people in need.

Overall, people were aware of such efforts made by the brands that they followed and were easily remembered during the conversation.

Regarding this perception, participants also reflected on their thoughts about the approach that some brands are taking towards supporting the community during the COVID-19 pandemic. People had a favorable opinion about brands not only joining the conversation but going "all in" in trying to mitigate some of the negative aspects of the pandemic. The interviewed participants used the words "favorable," "I'm glad," "it's good," and "I'm thankful" to describe their feelings towards brands' initiatives to help.

"I believe that... when this is over [the pandemic], people will know who were the brands that helped... I believe people will be aware of the brands that helped, or sent funds, or which what we consumed [bought] the company gave a part [of the profit] to people in need. I believe that this is something people will not forget and will have in mind who were the brands that helped the most during this phase" (Briana, 20).

During the interviews, the best ways that brands could help during this time of crisis were also discussed with participants. As the pandemic has progressed, people agree that it is no longer necessary to send informative content on the virus and how to prevent it. This information is already known and has been put into practice for many months. Examples of this include the importance of wearing a mask, staying at home, maintaining social distance, washing hands, not touching one's face, among others. Furthermore, the repetition by companies of such information at this point can generate stress, as one participant mentioned.

"I only see [campaigns] from all companies that say, "stay at home, stay at home, stay at home… [ ] generates a bit of an uncertainty, and a bit of fear" (Viviana, 57).

Most participants are certain about what types of support they expect from brands at this point of the pandemic. Suggestions include being transparent on the results of donations of funds and products to people in need, focusing on support for their employees, generating future employment opportunities, and supporting doctors and medical staff. However, one topic that was present in all of the conversations concerned the economic influence of the pandemic on people. Therefore, the top priority when expecting support from brands consisted of maintaining or lowering the prices of products and services, promotions, payment facilities, and credit. This can be found in many of the interview narratives with participants:

"It's good [to have low prices] because this is everyone's situation" (Abraham, 51).

"I think that [maintaining low prices] is a good strategy. There are a lot of people that are having a really hard time, economically speaking. There are many people that have been fired from their jobs, that don't have money to take food to their homes. The fact that companies are doing something for people, is something really good…[…]…they could try and lower prices for basic products" (Rocío, 46).

"[Companies should maintain] accessible prices, I know they probably also are being affected as a company, but consumers and citizens are also affected… some people had their salaries reduced, lost their jobs, they are not generating, or they depended on sales to have a commission, etc. So, I believe that the first thing companies should consider to help consumers is to maintain accessible prices" (Pablo, 23).

Additionally, this concern was also the number one priority with participants of the online survey as shown in Table 5.4. Following discounts, promotions, affordable shipping, and providing information about what they are doing with their staff and product management during the pandemic. Finally, several people emphasized that a good way to help consumers is to provide quick and easy ways to buy their products through friendly websites and apps. Because people at this moment cannot visit physical stores that often, it is important to include good quality pictures, videos, and other visual cues that can help consumers reduce uncertainty and ultimately make a decision.

Finally, it was also discussed how companies that are not supporting the COVID-19 cause are judged as not caring about people and the situation. However, participants commented on the fact that there is also a fine line between supporting the cause for the right reasons and supporting it only for publicity. In other words, people felt that some brands were not honest about their motives for providing support, rather, they believe it is only a marketing campaign designed to generate attention and attract positive consumer attention. As one participant mentioned:

TABLE 5.4 Suggested Brand Support to Consumers during the Pandemic

Brand Support
Reduce prices or maintain accessible prices
Offer discounts and promotions
Provide free or more affordable shipping
Provide additional information
Offer online transactions through website or app
Take sanitary precautions in stores
Maintain sufficient stock
Provide free or more accessible returns and changes for purchases
Facilitate the recycling of products
Provide drive-thru and contactless service
Make donations to people affected by COVID-19

Source: Own elaboration.

"It's good that companies want to contribute… but, if maybe they take it as an opportunity for them to promote themselves so that people notice them… to have more presence, to be more notable in social media…[ ] I don't think that's right. They should do it but not make publicity out of it…" (Pablo, 20).

5.5 DISCUSSION AND CONCLUSIONS

The objective of this research was to understand how consumers have modified their retail shopping behavior during the COVID-19 pandemic. The results surround the participants' experiences during the period of confinement, their adjusted shopping activities, and their perception of companies' reactions to the situation. Consistent with Mathur, Moschis, and Lee (2003), the findings suggest that when consumers transition into a new role or lifestyle to adapt to stressful life events, marketing opportunities are created. During this adaptation process, consumers rethink and reevaluate their consumption priorities, their need for products and services, and ultimately their brand preferences.

Thus far, the lessons learned from this research attempt to describe how the consumer has been influenced by the COVID-19 pandemic. First, consumers are informed and prepared. This is suggested by all of the participants reflecting on the fact that they are aware of news and developments about the virus itself every day, but also about what is happening with businesses, brands, and people around the world. Second, this crisis has caused consumers to also become more critical, as they frequently mentioned how they try to avoid fake news and rumors by researching information through credible sources and by questioning businesses' actions and real motivations. Third, consumers are more connected than ever before. Because most activities have been shifted to the online context, the time spent on the computer has increased dramatically. This prompts greater demand and higher expectations of online services by brands, such as friendly websites that are not only easy to navigate and search for products, but also address the problem of consumers' inability to physically see or touch the product. Fourth, consumers purchasing patterns have changed. The research suggests that people are concerned about the convenience of obtaining products, for example, good prices, better online shopping experiences, more accessible and quicker shipping, and show a preference for more personalized and local products. Most of the participants reflected on the fact that they had reevaluated their priorities and that their shopping had been focused on accessing necessary items rather than non-essential items.

Furthermore, as the results are analyzed, it can be suggested that consumers have experienced confinement and the overall pandemic with mixed emotions as the months go by. However, patterns among participants' reflections indicated that the emotions of the participants have undergone some shifts throughout the pandemic. Those emotions are consistent with other behaviors, such as the type of content they consume online and the purchases they make. Specifically, during the first months of confinement, fear was the predominant emotion, and more news and information about COVID-19 was consumed, as people wanted to be informed about the progression of the situation and find any indication about when we could resume normal activities. At this stage, panic purchases were more frequent and the findings are consistent with those of Yuen et al. (2020)—fear of the unknown, a lack of information, and social influence played an important role in motivating consumers to buy greater quantities of products. However, this research contributes to the panic purchasing literature by suggesting that consumers are not willing to accept that their abnormal purchases represent panic purchasing; rather, they consider them to simply be a matter of "being prepared."

As the days go on and the situation did not only not improve, but, in some places, worsened, fear turned into anxiety, causing consumers to cope by not obsessing over the news and finding other sources of information for entertainment and distraction. In terms of their purchasing behavior, once their basic needs were covered during the first stage, purchases for home offices and homeschooling were commonly made. This includes preparing a space or a room in the house appropriate for these purposes as well as buying other required products such as desks, chairs, computers, microphones, and earphones. Recently, we observe another shift in consumers' emotions, as they are now resigned to the situation and aware that it might take more time to finally control the disease. At this point, people are learning to live with new circumstances and taking actions to continue their lives. The behaviors that emerge from such emotions are finding content that helps people grow or advance personally and/or academically, such as courses, classes, and workshops. We also can observe many examples of people returning to abandoned hobbies or past interests, finding new interests, or focusing on pending renovations around the house. Figure 5.1 presents the timeline of consumer changes during the COVID-19 Pandemic.

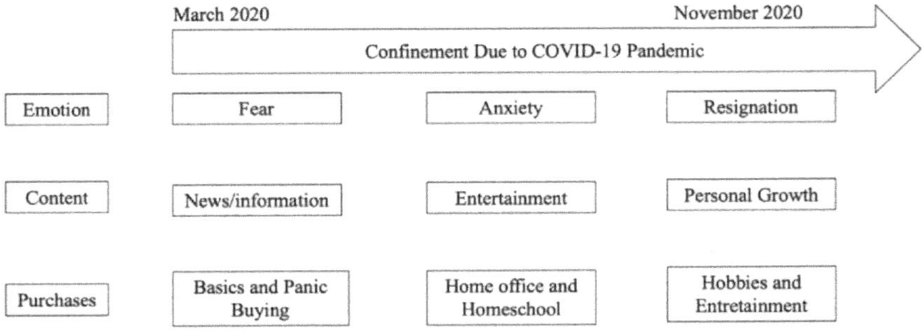

FIGURE 5.1 Consumer changes during the COVID-19 pandemic: Emotions, online content, and types of purchases.

Source: Own elaboration.

Since the emergence of the COVID-19 pandemic, brands had to quickly adapt and reinvent themselves to reach consumers. In a way, a company's products, services, and overall value proposition had to be modified to demonstrate that they were about much more than just profit—and care for people and the community. This research discovered that consumers were aware of the brand's actions during the emergency, and they have expectations derived from the psychological contracts they form with brands. Consistent with Sobande (2020), when a brand's actions and messaging reflect the idea that "we're all in this together," then such expectations are met and positive perceptions can be formed. However, this research also suggests that there is a very fine line between displaying real concern towards people and society and doing "good" just to pursue positive publicity and ultimately, profit. As this topic emerged during the interviews, it was clear that consumers are critical, educated, and informed when discussing brand activities.

It can be expected that even when the coronavirus disease gets under control and life returns to "normal," some of the changes in consumers will persist. Overall, the results of this research suggest that consumers are more informed, prepared, critical, connected, and prefer to relate to brands that seem caring, emphatic, and honest. Managers can benefit from these results by understanding the motivations prompting consumers to buy, their perceptions of brand communications during this crisis, and, more importantly, their expectations from brands during such complex times. Marketing communication campaigns can be created to exploit such propositions, considering that showing off efforts to help rather than delivering concrete action might result in negative perceptions and a switch of brand preferences (Kirk and Rifkin, 2020).

5.6 LIMITATIONS AND FUTURE RESEARCH LINES

No research is without limitations. First, this research presented the results of an exploratory study on the topic of consumer purchasing behavior during the COVID-19 pandemic, but by no means is its intention to generalize its findings. The participants selected, represent a sample of individuals belonging to a medium-high to high socioeconomic class in northern Mexico, for whom the pandemic conditions have been affected but are not as extreme as those of other people in the country. All of the participants have a home, food on their table, and the possibility to buy non-essential products and services. All interviewees had conserved their usual activities such as school or work. Therefore, their responses are expected to differ from people suffering from other conditions such as health issues or loss of employment, and, in those cases, their priorities are different. This research adopts only a qualitative approach, which was intended to understand the unique situation of each individual as well as the reasons underlying changes in their consumption habits. Further research can continue the conversation by including quantitative studies that address topics related to their consumption during this pandemic. For example, it would be interesting to create an experiment to analyze different company's support campaigns and test consumers' brand perceptions, willingness to support the cause, and ultimately their purchase intentions.

Additionally, the topic of panic purchasing was found to be particularly interesting in the findings. Future research could attempt to understand why consumers judge such behavior negatively in other people and yet do not recognize it in themselves, even though they have also engaged in it.

Overall, this research intends to shed light on how consumers modified their retail shopping behavior during the COVID-19 pandemic and calls for further research on this topic in the future.

KEYWORDS

- **consumer behavior**
- **marketing**
- **non-essential products**
- **novel coronavirus disease 2019 (COVID-19) pandemic**
- **panic buying**
- **shopping behavior**

REFERENCES

Ballantine, P. W., Zafar, S., & Parsons. A. G., (2014). Changes in retail shopping behavior in the aftermath of an earthquake. *The International Review of Retail, Distribution and Consumer Research, 24*(1), 1–13.

Forbes, (2020). *Las Marcas Ante el Coronavirus*. Red Forbes. Retrieved from: https://www.forbes.com.mx/las-marcas-ante-el-coronavirus (accessed on 18 February 2022).

Freeland, S., (2020). *Why People are Panic Buying and What Can Help*. UGA Today. Retrieved from: https://news.uga.edu/panic-buying-and-anxiety-tips/ (accessed on 18 February 2022).

Global Preparedness Monitoring Board, (2019). *A World at Risk: Annual Report of Global Preparedness for Health Emergencies*. Retrieved from: https://apps.who.int/gpmb/assets/annual_report/GPMB_annualreport_2019.pdf (accessed on 18 February 2022).

Kirk, C. P., & Rifkin. L. S., (2020). I'll trade you diamonds for toilet paper: Consumer reacting, coping and adapting behaviors in the COVID-19 pandemic. *Journal of Business Research*.

Littleton, T., (2020). *Empathy in Action: Brands Excelling in Empathy During the COVID-19 Pandemic*. The Social Element. Retrieved from: https://thesocialelement.agency/empathy-in-action-brands-excelling-in-empathy-during-the-covid-19-pandemic (accessed on 18 February 2022).

Loxton, M., Truskett, R., Scarf, B., Sindone, L., Baldry, G., & Zhao. Y., (2020). Consumer behavior during crises: Preliminary research on how coronavirus has manifested consumer panic buying, herd mentality, changing discretionary spending and the role of the media in influencing behavior. *Journal of Risk and Financial Management, 13*(8), 166.

Mathur, A., Moschis, G. P., & Lee. E., (2003). Life events and brand preference changes. *Journal of Consumer Behavior: An International Research Review, 3*(2), 129–141.

Montgomery, N. V., Raju, S., Desai, K. K., & Unnava. H. R., (2018). When good consumers turn bad: Psychological contract breach in committed brand relationships. *Journal of Consumer Psychology, 28*(3), 437–449.

Rogers, D., (2020). *Consumers are Putting Brands on Notice Over Coronavirus Behavior, Study Finds*. Retrieved from: https://www.campaignlive.co.uk/article/consumers-putting-brands-notice-coronavirus-behaviour-study-finds/1678821 (accessed on 18 February 2022).

Sobande, F., (2020). 'We're all in this together': Commodified notions of connection, care and community in brand responses to COVID-19. *European Journal of Cultural Studies, 23*(6), 1033–1037.

Yuen, K. F., Wang, X., Ma, F., & Li. K. X., (2020). The psychological causes of panic buying following a health crisis. *International Journal of Environmental Research and Public Health, 17*(10), 3513.

Zwanka, R. J., & Buff. C., (2020). COVID-19 generation: A conceptual framework of the consumer behavioral shifts to be caused by the COVID-19 pandemic. *Journal of International Consumer Marketing,* 1–10.

CHAPTER 6

Trust Me! Building Consumers' Trust During the COVID-19 Pandemic

FLOR MORTON

Department of Management, University of Monterey,
San Pedro Garza García, Mexico, E-mail: flor.morton@udem.edu

ABSTRACT

The appearance of the virus COVID-19 and its expeditious spread triggered a new and rapidly evolving situation for the whole world and has impacted consumers' in many different ways. Because of the pandemic and economic recession, consumers' perceptions of risk have changed, and trust has become more important in their buying decisions. Therefore, the present chapter analyzes some strategies to build consumers' trust in pandemic times by keeping consumers' safe, satisfying their current and post-pandemic needs, and engaging in CSR activities.

6.1 INTRODUCTION

In late December 2019, the appearance of the virus COVID-19 and its expeditious spread triggered a new and rapidly evolving situation for the whole world. In March 2020, the vast spread of the COVID-19 virus in the world qualified it as a global pandemic resulting in the introduction of legal regulations in most countries to slow down the spread such as social distancing, travel restrictions, use of mask, among others (Jeżewska-Zychowicz, Plichta, and Królak, 2020). The impact of the current pandemic on the global economy is a phenomenon not seen since the 1930s Great

Marketing by Contingency in the Time of COVID-19: Overcoming Business Crises and Meeting Marketing Challenges. Alicia De La Peña, PhD & Juan Bernardo Amezcua Nunez, PhD (Eds.)

Depression and it has been described by some authors as a 'Black Swan Event' referring to a striking event that changes the world and that is characterized by being unpredictable and having a massive impact. The short-term impact of the pandemic has been immediate and effortlessly felt. However, the long-term and profound economic, political, social, and cultural impacts are yet to be seen. COVID-19 will surely change how we see the world, how we think and our behavior (He and Harris, 2020; Mazzoleni, Turchetti, and Ambrosino, 2020; Taleb, 2008).

The pandemic surprised consumers in the developed world with the idea that their basic needs might not be met; this was a shock for consumers that before the pandemic took for granted that their needs could be easily satisfied by the wide availability and overload of choices of products and services. Also, the pandemic has given the opportunity for consumers to reflect on the basic meaning of consumption and its impact not only on themselves, but on society and on the environment as well. The Organization for Economic Co-operation and Development (OECD) has made a call to avoid a return to 'business as usual' and rather develop policies to 'build back better' for a more sustainable, resilient, and inclusive economic recovery from the COVID-19 crisis (OECD, 2020). Consequently, the COVID-19 pandemic will probably represent one of the most significant environmental changes in modern marketing history, with important implications on basic marketing philosophy and strategy, corporate social responsibility (CSR), and consumer ethics (He and Harris, 2020).

Marketing literature recognizes customer loyalty and trust as key factors for long-term companies' profitability and growth (Aslam, Hussain, Farhat, and Arif, 2020). In a recent webinar of the American Marketing Association, the role of trust was highlighted as becoming more important now than before the pandemic in building a strong brand (Jestice, 2020). Despite trust has been a concept that has received considerable attention in the marketing literature (Moorman, Zaltman, and Deshpandé, 1993; Sirdeshmukh, Singh, and Sabol, 2015; Urban, Sultan, and Qualls, 2000), there is scant literature that analyzes the concept of trust under the pandemic context. Therefore, the present chapter is structured as follows. First, it reviews the concept of trust and its dimensions; second, it examines the trust challenges amid the pandemic; third, it offers some thoughts on strategies to build consumer's trust in the COVID-19 context; and finally, it presents some conclusions and research questions for future research.

6.2 CONSUMER'S TRUST

Previous authors have been particularly interested in consumers' trust, because it is considered an essential element for fostering strong relationships, loyalty, and sustainable market share (Flavian, Guinaliu, and Gurrea, 2006; Moorman, Zaltman, and Deshpandé, 1993; Urban, Sultan, and Qualls, 2000). However, trust is perhaps one of the terms whose concept is hardly agreed within diverse academic disciplines. In marketing, trust is generally understood as consumer's perceived reliability on the company and/or the company's brand or products (Flavian and Guinaliu, 2006).

On one hand, trust has been conceptualized as a belief, confidence, or expectation about the trustworthiness of an exchange partner as a result of the partner's expertise, reliability, and integrity (Anderson and Weitz, 1989; Dwyer, Shurr, and Oh, 1987). On the other hand, this concept has been defined from a behavioral point of view, considering trust as an intention or behavior that reflects confidence on a partner which involves a certain degree of vulnerability and uncertainty on the part of the trustor (Moorman, Zaltman, and Deshpande, 1992; Zand, 1972).

Most research has focused on the consequences of trust, and relatively few studies have analyzed company behaviors, practices, or mechanisms that contribute to building consumer's trust (Sirdeshmukh, Singh, and Sabol, 2015). According to Moorman, Zaltman, and Deshpande (1992), belief, and behavioral intention must be present for trust to exist. The authors extend the psychological view to include a sociological component to the definition of trust as a willingness to rely (sociological component) on an exchange partner, such as a brand or company, in whom one (e.g., a consumer) has confidence (psychological component). Confidence and reliance, in turn, indicate that both vulnerability and uncertainty are critical to trust. On one hand, a certain degree of vulnerability is necessary because it implies that outcomes of the exchange relationship are consequential for the trustor and therefore trust becomes necessary. On the other hand, uncertainty is critical because trust would be unnecessary if the trustor can control or has a complete knowledge about an exchange partner's actions (Moorman, Zaltman, and Deshpande, 1993). For instance, making a purchase represents a certain level of trust as consumers have some degree of uncertainty about the products' quality and safety as well as about the company's response in case of product or delivery failures which represent potential negative consequences.

Furthermore, when consumers have a certain degree of perceived risk in their decision-making behaviors, they can avoid certain decisions such as product purchases, due to the uncertainty and potentially negative consequences (Shin and Kang, 2020).

Sirdeshmukh, Singh, and Sabol (2015) define trust as consumer's expectations on the company's reliability to deliver its promises, these authors propose a multidimensional and multifaceted model for the behavioral components of trustworthiness in consumer-firm exchange relationships offering insights into trust-building and trust-depletion processes. This model identifies operational competence (expectation of a company's ability of consistent and competent performance to fulfill promises made to the consumers), operational benevolence (behaviors that reflect a company's motivation to place consumers' interests ahead of its own with a sincere concern for their wellbeing), and problem-solving orientation (consumer's evaluations of the company's motivations to anticipate and solve problems) as dimensions of consumer's trust. Furthermore, the authors propose that in a service context, consumer's trust develops around two facets: front line employees (FLEs) and management policies and practices (MPPs); it is likely that consumers make independent judgments for these two facets during the course of a service exchange.

Previous literature on consumer's trust in e-commerce has proposed that trust in an online store is a key driver for sales and is determined by three dimensions: competence, integrity (suggests that a company acts in a reliable and honest manner), and benevolence (Chen and Dhillon, 2003). Other research has proposed a model for consumer's trust in internet shopping identifying that trustworthiness of Internet merchant (ability, integrity, and benevolence), trustworthiness of Internet shopping medium (technical competence, reliability, and medium understanding), contextual factors (effectiveness of security infrastructure), individual's trust propensity, and other factors (such as company size and demographic variables), are main antecedent influences on consumer trust in Internet shopping (Lee and Turban, 2001).

In summary, trust is considered a multidimensional construct and recurring dimensions identified by the literature on this topic are benevolence, competence, and integrity. Based on previous literature, in this chapter, trust is understood as an intention or behavior that reflects consumer's willingness to rely on a company despite the presence of a certain degree of uncertainty and vulnerability because the company or brand is perceived as competent, benevolent, and with integrity. Table 6.1 identifies dimensions and outcomes of trust based on the literature review presented in this section.

TABLE 6.1 Dimensions and Outcomes of Trust

Trust Dimensions	Trust Outcomes
Benevolence	Purchase
Competence	Strong relationships
Integrity	Loyalty
	Sustainable market share

Source: Developed by the author based on previous literature.

6.2.1 COVID-19 AND CONSUMER'S TRUST

It could be expected that some contexts, such as an economic crisis or a threat to consumers' health and wellbeing have the potential to impact consumers' perceptions of risk and make them more risk-averse in their decisions; as a result, consumers' trust in a particular company or brand becomes particularly important in such contexts. For instance, COVID-19 pandemic has definitely raised companies' awareness of the need to foster consumer's trust in order to survive the current crisis (Edelman Trust barometer, 2020).

On March 11 2020 the disease COVID-19 was declared a pandemic by the World Health Organization (WHO) (Forbes, 2020). This disease was caused by the SARS-CoV-2 new coronavirus, and it is considered extremely contagious and mortal for certain vulnerable groups such as elder people, pregnant women, and those with other diseases such as diabetes, hypertension, among others (Gobierno de Mexico, 2020). The vast spread of the virus in the world required the introduction of legal regulations in most countries to slow down the spread, including a lockdown, social distancing, travel restrictions, use of mask, among others (Jeżewska-Zychowicz, Plichta, and Królak, 2020).

Following the announcement of the lockdown, an unusual retail consumer behavior was reported all over the world, and one of the most immediate effects of the pandemic was the apparent supply shortages of certain products caused by unanticipated spikes in demand. In the early stages of the COVID-19 dramatic images of supermarket shelves emptied of key items such as food, bottled water, toilet paper, hand sanitizers, hand soap, among products, as people began to make panic purchases for these products (Laato, Islam, Farooq, and Dhir, 2020; Mohan, Berg, and Poblet, 2020). Despite the emphasis of government officials and industry representatives of a sufficient supply of these products, short-run panic buying behaviors became self-perpetuating in response to the fear of a global disruption of production and supply chains (Hobbs, 2020).

An economic recession came along with the pandemic affecting countries all over the world. Previous literature recognizes important effects of crisis on consumers both psychologically and financially. In a crisis, there is a negative mental impact on consumers making them more eager to save money because they lost their job or they fear to lose it; financially, as a result of a change in perception of risk, consumers reduce consumption by prioritizing necessary products and switching to cheaper options (Köksal, and Özgül, 2007; Mogaji, 2020; Sharma and Sonwalker, 2013). Because of this change in perception of risk, it is fundamental for brands to build and/or increase consumers' trust in their products and value propositions to protect and maintain the relationships with their current customers and create new ones with potential buyers. Furthermore, the need for social distancing has accelerated the digital transformation of companies and people are now spending increased periods of time online. In addition, unemployment rates have also increased, as well as the rate of cyber-attacks as some people have turned to cyber-crime to support themselves during this time of crises (Tran, 2020), creating new considerations around the safe and ethical use of data and technology.

The COVID-19 pandemic emerged as an unpredictable event that has been a shock with major effects to our collective systems; major shifts in consumer traits, buying patterns, global interconnectedness, and psychographic behavior are taking place in the short, medium, and long term (Zwanka and Buff, 2020). Despite the shock caused by the pandemic, it has also been considered as a catalyst to rebuild trust, because trust in the current economic system has declined in part due to an increase in disinformation, polarization, rapid technological change, among other reasons. In spite, companies cannot control the pandemic situation, they can work on being intentional about trust and balancing stakeholder trust during and after this crisis, if done well, this could have the potential to accelerate a company's recovery and to emerge stronger and maintain business continuity (Lee et al., 2020). Therefore, this chapter analyzes some strategies to build consumers' trust in times of pandemic.

6.3 BUILDING TRUST AMID THE PANDEMIC

The rapid spread of COVID-19 and its consequences in the market environment required organizations to develop strategies in an entrepreneurial agility and flexibility mode, almost to the point of hypermobility in the development of systems, operations, and tactics to reach consumers (He and Harris, 2020). It is relevant that companies consider consumers' trust in the development and implementation of those strategies and procedures.

According to the most recent report of Edelman Trust Barometer (2020), there is a higher need for brand trust today than in the past, results from this report show that 61% of people appreciate when brands spend money to keep consumers informed about the solutions and help provided to others by the brand or company during the pandemic. Additionally, results suggest that trust builds loyalty, engagement, and advocacy, and 89% of the respondents considered that brands must allocate money and resources to create products that help people overcome pandemic-related challenges in order to earn or keep their trust. Furthermore, the Edelman report reveals that people identify a bigger need for trust in brands because of both personal and societal reasons. Personal reasons include people's identification of an increase in personal vulnerability as brands are collecting more personal information and consumers are struggling financially and can't afford to waste money. Reliance on brands to keep the community safe, help consumers get through day-to-day challenges and do important things for them, and because consumers use brands to express their values and they rely on brands to be a good symbol. Societal reasons include brand's environmental impact as a result of their production and delivery processes, brand's involvement in major social issues and problems, and the potential of technology innovations such as AI and robotics to harm society if misused.

6.3.1 BUILDING TRUST THROUGH BENEVOLENCE: KEEPING CONSUMERS SAFE

One of the major challenges faced by companies with the reopening of activities is how to keep consumers safe and reduce perceived risk both offline and online. Companies must improve perceptions of safety by the implementation of both contagion-risk and cybersecurity-risk reduction strategies, but with a sincere motivation to protect consumer's wellbeing and not only because of legal regulations.

6.3.1.1 CONTAGION-RISK REDUCTION STRATEGIES

For instance, in order to attract people to visit hotels, most branded hotels such as Marriott, Hilton, and Hyatt are now implementing technology innovations that reduce guest interactions with hotel employees, these include mobile and kiosk check-in systems, robot cleaning systems and/or an update of their cleaning procedures (e.g., electrostatic sprayers and ultraviolet-light technology) (Shin and Kang, 2020).

The grocery retail industry has encountered unique challenges during the pandemic, which led to various transformations including changes in consumer perceptions as they have higher expectations for in-store safety and behavior as their shopping frequency and in-store duration have reduced; and they now spend more per shopping trip (Wang, Xu, Schwartz, Ghosh, and Chen, 2020). Restaurants have seen a decrease in visits, instead people are increasingly ordering food through delivery apps (Gutiérrez, 2020). However, these apps charge high percentages of sales commissions to restaurants making them increase their prices up to 30–40% or reduce their margins (Expansión, 2020). Therefore, restaurants are trying to motivate consumers to order by phone and pick up their food or visit the restaurant where they have implemented some actions to reduce contagion risk like temperature control at the entrance, sanitizer mats, digital menus, utensils wrapped in plastic bags, among others.

The fashion retail industry has also faced its challenges; previous research has found that people perceive contamination of clothes touched by others, and this can influence their willingness to buy clothes (Argo, Dahl, and Morales, 2006). Therefore, fashion stores have been forced to innovate along the fashion value chain implementing 3-D design, virtual sampling, e-commerce, among other innovations (Alves, Ribeiro, and Saravanan, 2020). Also, short-term strategies have been implemented, such as temporarily closing fitting rooms accompanied by changes in product return policies, including extension of return term limit and the option of online returns via postal services. In general, both the retail and customer service industries have implemented similar mandatory measures to reduce the risk of contagion, such as the obligation to use face masks, temperature control, antibacterial for hands and sanitizing mats at the entrance, certain opening hours, entrance prohibition to vulnerable customers, among other contagion risk prevention actions. The challenge faced by companies in these industries is how to transmit a contagion risk free environment while still delivering a gratifying shopping experience. Also, supervision of the correct implementation and supervision of these contagion risk reduction strategies is crucial for the company to transmit a genuine interest to keep their consumers safe.

6.3.1.2 CYBERSECURITY-RISK REDUCTION STRATEGIES

As previously mentioned, the need for social distancing has accelerated the digital transformation of companies. However, this has imposed some challenges to trust, such as cyber risks, including the detection and prevention of fraud, misuse of technology and data, and privacy incidents. This matter has become crucial and challenging, particularly for those companies that have not been considered by their leaders as "tech companies" and have not made considerations of cybersecurity as much as is required now that their work and interactions are predominantly virtual. Since data security and privacy are important issues for trust, cybersecurity is a key element to build trust among companies' stakeholders and drive sustainable success (Lee et al., 2020); hence, companies need to protect the cybersecurity, not only of their consumers, but across their key stakeholder network (suppliers, partners, customers, employees, etc.), as well. For instance, the Internet has given companies access, at a relatively low cost, to an indefinite number of consumers and e-commerce platforms allow firms to directly sell products to consumers resulting in increased sales (Lee et al., 2018). E-commerce platforms collect and track customer information for data-driven marketing efforts, but unfortunately some of these platforms do not provide an adequate safeguard for this data and security risks such as personal information leakage have negative effects on online transactions (Bandara et al., 2020; Hubert et al., 2017; Tran, 2020).

Consumers' loyalty to a website has been found to be closely linked to the levels of trust and trust is particularly influenced by the individuals' perceived security regarding the handling of their private data (Flavián and Guinalíu, 2006). Similarly, Aslam, Hussain, Farhat, and Arif (2020) studied the factors that influence consumers' trust and its subsequent impact on loyalty. According to the authors, in developing countries where the majority of e-customers are considered to be risk avoiders, trust becomes exceedingly imperative in shaping the attitude of customers toward e-commerce; in their study user interface quality, perceived privacy, information quality, perceived security risk, and e-commerce awareness were found to be determinants of customers trust in e-commerce and trust was found to be a predictor of loyalty. In a pandemic condition, high levels of fear motivate consumers to rely more on their perceived effectiveness of e-commerce platforms (i.e., consumers' perceptions that a particular e-commerce platform has safeguard mechanisms to protect online transactions from potential risks). Therefore, as recommended by previous literature, firms should seek opportunities to provide consumers effective online safeguards including privacy protection such as external payment methods for secure online transactions where the user's data is encrypted (e.g., PayPal) and credit card guarantees with financial institutions to protect and compensate online consumers in case of fraud (Tran, 2020). Moreover, it could be helpful in building trust if companies communicate clearly the strategies they implement to protect their customers and the solutions or support offered in case of a cyber fraud or privacy incident.

6.3.2 BUILDING TRUST THROUGH COMPETENCE: HELPING CONSUMERS MEET THEIR NEEDS

The pandemic surprised consumers in the developed world with the idea that their basic needs might not be met and many consumers made panic purchases in response to the fear of a global disruption of production and supply chains (Hobbs, 2020). Supply for retail and consumer services has suffered from the present unusual situation in several ways. With many companies implementing a distribution system built around just-in-time manufacturing and delivery, sudden spikes in the demand for their products caused by panic buying created short-run stockouts (Mussell, Bilyea, and Hedley, 2020). However, Guerrieri, Lorenzoni, Straub, and Werning (2020) argue that the effects of the pandemic on supply disruptions could cause supply shocks that initially may only affect a single industry, but can spread quickly into other industries causing less availability and increases in price of products, making consumption more high-priced and thus also less attractive.

Despite tempting, cutting costs can be a fatal move, especially if the costs that are cut reduce the company's ability to consistently satisfy the needs of consumers and result in a deterioration in customer service, absence or inefficient FLEs, delayed delivery times or a lower quality product or service. Eliminating intermediaries to reduce costs could be an adequate strategy, but it should be taken carefully ensuring the products availability for customers and considering the long-term consequences of reducing intermediaries. For instance, now firms can use their website and social media as sales channels reducing intermediaries and making it cheaper for the company as well as providing an easier way for the customer to reach information about the company or product. Nevertheless, it is well known that a disadvantage of e-commerce is that consumers cannot physically examine products and have to rely only on information obtained through product presentation formats such as pictures, written descriptions, videos, among others, which can lead to product uncertainty (Morton, 2018; Song et al., 2020; Tran, 2020). This imposes the challenge of finding innovative ways to transmit product information and characteristics that convey information obtained by directly observing, touching, smelling, and/or tasting the product to help consumers evaluate the product and diminish the produced uncertainty of not being able to physically examine it. Examples of this include Ray Ban's virtual mirror, Nike augmented reality try on app, and research developments on digital smell and taste technologies (Cheok and Karunanayaka, 2018).

Additionally, in times of crisis, new technologies can have a great impact on marketing innovations. According to a study by Köksal and Özgül (2007), despite the significant financial limitations imposed by a crisis, increasing the research and development budget has an important impact on performance. The authors suggest that technology and production methods that can help reduce costs should be increased along with the development of products that capture niche markets. Furthermore, it is important for marketers to think carefully about segmentation and targeting. The pandemic has come to change people's needs; hence, as a company's consumers' needs change, the relevance of a company's product or service might change as well. Therefore, companies should have a clear target segment and focus on being empathic by understanding and meeting their current and post-pandemic needs, by helping them deal with the realities of the pandemic and anticipate future needs. Additionally, companies could also target current offers to segments that truly value them and clearly ensure differentiation emphasizing how the company's products are still essential for them and how they can be beneficial to their lives.

Consumers become more price-conscious and price-driven in times of crisis. Hence, during tough economic conditions, companies might see price as a quick fix or perhaps the only way to add value for their customers (Piercy, Cravens, and Lane, 2010). As the competition for the demand becomes fierce, companies may be tempted to lower their prices and start a price war as a desperate attempt to boost sales. Although lowering prices could be effective in increasing sales, it can reduce profit margins to dangerous low levels. When companies lower their prices, they are making the assumption that their target segments are price buyers. However, not all consumers buy exclusively based on price but on perceived value as well (De Toni, Milan, Saciloto, and Larentis, 2017).

According to Ang et al. (2000), product quality and pricing decisions should always be made together. Turbulent economic times accentuate the need to examine and innovate the price strategy of the company. To survive and recover from the pandemic, such innovations should be made with a clear focus on meeting consumers' value requirements, for instance, offering higher quality products at the same price has shown to have a positive effect on performance (Köksal and Özgül, 2007). It is also important for companies to avoid price strategies that have been shown that represent a risk in building or maintaining consumers' trust, such as dynamic pricing (adjusting prices based on demand). For example, geographically, and temporarily price discriminations have been effective in the past. These types of price discriminations rely on consumer's unawareness, in the Internet, they become more difficult to implement because of the borderless flow of information that results from the ease of search and consumer interactions that can raise awareness of such price discrimination. Yet, consumers accept price variations based on differences in costs or on social norms (e.g., different product prices for students) (Garbarino and Lee, 2003). Therefore, firms could emphasize savings and offer price benefits when consumers use their e-commerce platforms because of the relatively lower costs of this channel compared to those of traditional stores (Tran, 2020).

Delivering superior value during and after a crisis is a major challenge for companies and pricing decisions can have long-term implications; hence, moving from tactical pricing (short-term) to strategic pricing (long-term) decisions should be a prime criterion in order to reach consumers who are willing to pay the price for the product or service that meets their value requirements (Piercy, Cravens, and Lane, 2010). Because pricing is a key element in the revenue recovery strategy of a company, the current pandemic demands pricing leaders to be creative in meeting consumers' needs while still preserving product value and ensuring competent performance.

6.3.3 BUILDING TRUST THROUGH INTEGRITY: CORPORATE SOCIAL RESPONSIBILITY (CSR)

Undoubtedly, the current crisis has put companies under test for their commitment to corporate social responsibility (CSR), ethical conduct, and integrity. Currently, there is an increased demand for CSR activities in response to the pandemic. It has been reported that some companies have tried profiteering from this crisis by inflating prices or making misleading claims about products. Fortunately, many companies not only have avoided unethical business practice, but have also leveraged the opportunities COVID-19 offers from a social perspective and engaged in various CSR activities the number of CSR activities in response to the crisis is too long to enumerate in this chapter, but this section offers some examples. For instance, some companies used their factories to produce personal protective equipment, ventilators, hand sanitizer, among other products that could offer immediate help and assistance to fight against the virus; and some of these companies, instead of selling, donated these products. Supermarkets designated specific opening hours for elder people, donated food to charitable organizations, banks waived interest and extended pay time periods, among other activities (He and Harris, 2020; Manuel and Herron, 2020).

Previous literature recognizes that a crisis can affect companies in many different ways and if not managed correctly, it can have severe consequences for the organization in the long term. An inadequate demand for products and services combined with a fierce competition in the market environment force some companies to reduce production, which leads to employee layoffs or disruption in the supply chain that can damage the organization since it puts to test the reputation and trust with stakeholders (Ghandour and Benwell, 2012; He and Ran, 2015; Köksal and Özgul, 2007). The empathy and commitment shown by some companies to treat their employees well by making them feel secure in their jobs with actions such as maintaining or increasing wages, helping them to adequate their homes with office supplies and furniture, monitoring their health condition and communicating transparently about current infection status among employees is a form of internal CSR that could also help build and restore both employees trust, engagement, and performance (Jia, Yan, Liu, and Huang, 2019; Lee et al., 2020); since employees are an important source for transmitting trust to consumers, building or increasing their trust, engagement, and performance can also impact consumer's trust.

Additionally, despite a strong call for global unity, solidarity, and cooperation, the current situation has already caused significant geopolitical tensions, manifestations of these tensions include the surge of nationalist sentiments of consumer behavior and negative feelings toward products from a specific country. This puts to test global supply systems as the disruption of the global supply chain forces and encourages consumers to buy local products if available (He and Harris, 2020). However, companies can view this as an opportunity to engage in CSR activities helping local companies, especially those of vulnerable groups, by including them as part of their supply chain.

An interesting phenomenon known as coopetition is taking place amid the pandemic (Crick and Crick, 2020). Coopetition refers to the cooperation of industry rivals (Brandenburger and Nalebuff, 1996). For instance, a group of cybersecurity experts from rival tech companies including senior professionals from Microsoft and Amazon with the name of "COVID-19 CTI League" have come together to combat cybersecurity threats associated with the pandemic (Security, 2020). Coopetition can also take the form of showing empathy and solidarity to rivals. For example, an interesting proof of being a trustful company is that shown by the recent text published by Burguer King on their social media asking consumers to buy McDonalds' hamburgers (Forbes, 2020).

The COVID-19 pandemic forced people around the world to stay at home, this situation has resulted in a high prevalence of psychological distress, which was intensified by the flow of information about the rapid increase in mortality, the limited availability of diagnostic tests, treatment, cure, as well changes in food availability, making people experience fear and making purchasing decisions based on panic. Recent literature has attributed this particular behavior to several reasons such as cyberchondria caused by an overload of information about COVID-19 (Laato, Islam, Farooq, and Dhir, 2020). As consumers are paying more attention to authoritative sources and some companies can be considered as such, another form of CSR is using marketing to minimize the impact of consumers' over-acquisition of goods during crises. This firm-driven approach is centered on the new concept of consumer-centric socially responsible consumption (CCSRC) is based on the well-being marketing philosophy and expands the concept of socially responsible consumption to include consumer's responsibility to other consumers. This concept suggests that before panic buying begins, companies have the opportunity to focus consumers' attention on the needs of other customers and diminish the negative impact of this behavior (Anderson and Anderson, 2020). This approach could positively impact consumers' perceptions of a company's benevolence and integrity.

6.4 CONCLUSION

COVID-19 emerged as a Black Swan event with major changes in consumers' behavior, lifestyle, habits, perceptions, and needs (Zwanka and Buff, 2020) and has exposed businesses' vulnerability to extraordinary external forces. The pandemic has also given the opportunity for consumers to reflect on the basic meaning of consumption and its impact. A company's genuine and authentic CSR could build more trust among society and its customers. In fact, some argue that the bond established by brands with their customers during this crisis could be more meaningful and lasting than before de pandemic (He and Harrris, 2020). Therefore, companies should focus more on their efforts on contributing to solving social and environmental issues during the pandemic as well as in the post-pandemic period. Consequently, the COVID-19 pandemic will probably represent one of the most significant environmental changes in modern marketing history.

Because now consumers perceive a higher vulnerability and risk, trust has become more important now than in previous pandemic times in building strong brands (Jestice, 2020). Considering the impact of trust on consumers loyalty and its implication for profitability and growth, understanding, and being intentional in building trust during and after this crisis could have the potential to accelerate a company's recovery and to emerge stronger and maintain business continuity (Lee et al., 2020).

The economic crisis puts financial pressures on firms to cut costs. It is known that during crises, promotion efforts are cut because they do not show tangible and immediate results (DeDee and Vorhies, 1998). However, literature suggests that the promotion strategies during a crisis are of great importance to determine a company's performance after the crisis (Kim, 1992; Werner, 1991). When consumers experience a decrease in their purchasing power they tend to shop more rationally (Köksal and Özgül, 2007); hence, advertising campaigns during and after the pandemic should focus on communicating rational motives as reliability, safety, and durability of a company's product to reduce perceived risks. Also, ensuring an open, transparent, empathic, and honest communication with consumers is a way to build trust, companies should not stop communicating with their customers because it is important to keep consumers informed on how the company is being impacted, which measures the company is implementing to protect consumers' wellbeing, how the company's products can satisfy their needs and announce new products for future needs, as well as to communicate CSR efforts.

This chapter presented some challenges faced by companies as a result of COVID-19, it was not the intent for this analysis to be exhaustive, but to offer some initial thoughts on how to build trust through strategies that could impact consumer's perceptions of a company's competence, benevolence, and integrity during the pandemic and recession context. It is important to learn from the current situation both to help brands and companies to deal with the ongoing COVID-19 pandemic and to be able to react better to similar situations in the future (Laato, Islam, Farooq, and Dhir, 2020). Finally, some research questions on consumers' trust derived from the pandemic are suggested for future exploratory and/or explanatory research. How does a company's CSR response in the pandemic impact both the general public and consumers' trust on the brand? Does helping consumers to reduce their perceived risk or fear in a crisis can increase their trust in our brands? Does product uncertainty that surges in e-commerce platforms intensifies because of changes in perceptions of risk caused by a pandemic? Which price strategies are more effective for recovery from a recession without damaging consumer's trust? Do certain types of CSR responses during a pandemic are more effective than others? Which strategies to keep consumers' safe have more impact on trust?

KEYWORDS

- **consumer trust**
- **corporate social responsibility**
- **COVID-19**
- **pandemic**

REFERENCES

Alves, S. R., Ribeiro, H., & Saravanan. R., (2020). The digital transformation of traditional fashion retail for sustainable growth. *Economic and Social Development: Book of Proceedings,* 236–245.

Anderson, E., & Weitz. B., (1989). Determinants of continuity in conventional industrial channel dyads. *Marketing Science, 8*(4), 310–323.

Anderson, S., & Anderson. J., (2020). Leave some for me! The role of marketing in influencing responsible consumption during times of crisis. *Journal of Strategic Marketing,* 1–19.

Argo, J. J., Dahl, D. W., & Morales. A. C., (2006). Consumer contamination: How consumers react to products touched by others. *Journal of Marketing, 70*(2), 81–94.

Aslam, W., Hussain, A., Farhat, K., & Arif. I., (2020). Underlying factors influencing consumers' trust and loyalty in E-commerce. *Business Perspectives and Research, 8*(2), 186–204.

Bandara, R., Fernando, M., & Akter. S., (2020). Explicating the privacy paradox: A qualitative inquiry of online shopping consumers. *Journal of Retailing Consumer Services, 52*(101947), 1–9.

Brandenburger, A. M., & Nalebuff. B. J., (2011). *Co-Opetition.* New York, NY: Doubleday Dell Publishing Group Inc.

Chen, S. C., & Dhillon. G. S., (2003). Interpreting dimensions of consumer trust in E-commerce. *Information Technology and Management, 4*(2), 303–318.

Cheok, A. D., & Karunanayaka. K., (2018). *Virtual Taste and Smell Technologies for Multisensory Internet and Virtual Reality.* Cham: Springer.

Crick, J. M., & Crick. D., (2020). Coopetition and COVID-19: Collaborative business-to-business marketing strategies in a pandemic crisis. *Industrial Marketing Management, 88*, 206–213.

De Toni, D., Milan, G. S., Saciloto, E. B., & Larentis. F., (2017). Pricing strategies and levels and their impact on corporate profitability. *Revista de Administração (São Paulo), 52*(2), 120–133.

DeDee, K. J., & Vorhies. D. W., (1998). Retrenchment activities of small firms during economic downturn: An empirical investigation. *Journal of Small Business Management, 36*(3), 46–61.

Dwyer, F. R., Schurr, P. H., & Oh. S., (1987). Developing buyer-seller relationships. *Journal of Marketing, 51*(2), 11–27.

Edelman, (2020). *Trust Barometer Special Report: Brand trust in 2020.* Retrieved from: https://www.edelman.com/sites/g/files/aatuss191/files/2020-06/2020%20Edelman%20Trust%20Barometer%20Specl%20Rept%20Brand%20Trust%20in%202020.pdf (accessed on 18 February 2022).

Expansión, (2020). México. *Las Apps de Delivery, ¿Amigas o Enemigas de Los Restaurantes?* Retrieved from: https://expansion.mx/tecnologia/2020/07/14/las-apps-de-delivery-amigas-o-enemigas-de-los-restaurantes (accessed on 18 February 2022).

Flavián, C., & Guinalíu. M., (2006). Consumer trust, perceived security and privacy policy. *Industrial Management & Data Systems, 106*(5), 601–620.

Flavian, C., Guinaliu, M., & Gurrea. R., (2006). The role played by perceived usability, satisfaction and consumer trust on website loyalty. *Information & Management, 43*, 1–14.

Forbes, (2020). *Burger King Pide Comprar a McDonald's en Apoyo por Pandemia.* Retrieved from: https://www.forbes.com.mx/negocios-burger-king-mcdonalds-apoyo-pandemia/ (accessed on 18 February 2022).

Forbes, (2020). *OMS Declara Pandemia por el Coronavirus COVID-19.* Retrieved from: https://www.forbes.com.mx/mundo-pandemia-coronavirus-covid-19/ (accessed on 18 February 2022).

Garbarino, E., & Lee. O. F., (2003). Dynamic pricing in internet retail: Effects on consumer trust. *Psychology & Marketing, 20*(6), 495–513.

Ghandour, A., & Benwell. G., (2012). A framework of business recovery in the aftermath of a disaster. *International Journal of Business Continuity and Risk Management, 3*(3), 263–274.

Gobierno De México, (2020). *Criteria for Populations in a Situation of Vulnerability that may Develop a Complication or Die from COVID-19 in the Reopening of Economic Activities in Work Centers.* Retrieved from: https://coronavirus.gob.mx/wp-content/uploads/2020/07/Criterios_vulnerabilidad_27Julio2020.pdf (accessed on 18 February 2022).

Guerrieri, V., Lorenzoni, G., Straub, L., & Werning. I., (2020). *Macroeconomic Implications of COVID-19: Can Negative Supply Shocks Cause Demand Shortages?* National Bureau of Economic Research. (No. w26918).

Gutiérrez, A. L., (2020). *COVID-19 y Desempleo Suben Demanda de Aplicaciones de Comida a Domicilio.* El Financiero. Retrieved from: https://www.elfinanciero.com.mx/empresas/covid-19-y-desempleo-suben-demanda-de-aplicaciones-de-comida-a-domicilio (accessed on 18 February 2022).

He, H., & Harris. L., (2020). The impact of covid-19 pandemic on corporate social responsibility and marketing philosophy. *Journal of Business Research, 116*(2020), 176–182.

He, L., & Ran. Y., (2015). The correlation of brand equity and crisis: A review and directions for future research. *Modern Economy, 6*(03), 392–397.

Hobbs, J. E., (2020). Food supply chains during the COVID-19 pandemic. *Canadian Journal of Agricultural Economics/Revue Canadienne D'agroeconomie,* 1–6.

Hong, I. B., & Cho. H., (2011). The impact of consumer trust on attitudinal loyalty and purchase intentions in B2C e-marketplaces: Intermediary trust vs. seller trust. *International Journal of Information Management, 31*(5), 469–479.

Hubert, M., Blut, M., Brock, C., Backhaus, C., & Eberhardt. T., (2017). Acceptance of smartphone-based mobile shopping: Mobile benefits, customer characteristics, perceived risks, and the impact of application context. *Psychology & Marketing, 34*(2), 175–194.

Jestice, G., (2020). *Building Brand Trust: The Secret to Thriving in the New Normal [Webinar].* AMA Marketing. https://www.lucidpress.com/pages/resources/webinars/brand-trust?utm_source=marketo&utm_medium=email&utm_campaign=ama-2020-07&utm_content=e1+post-event-email-full&utm_grouping=post-event&mkt_tok=eyJpIjoiTjJNeU1ERXdOV0k0TW1FMCIsInQiOiI4dFQ4MEx5NFA4N25UOExmYzFOX C9pc1dpYmF6bjlqZkRiV2V6cVwvZDdmU1RJMjZSTEJveU85XC9hY1wvYUVMXC9HTG5OaytUVndcL1lxSldtSHJsc2JIWmlEU1o3VU04UHJ0dk5ZOHNtMFZjeVZ2elwvYUlXbDBURlFycXpNQTEzQ3htTUQ ifQ%3D%3D (accessed on 18 February 2022).

Jeżewska-Zychowicz, M., Plichta, M., & Królak. M., (2020). Consumers' fears regarding food availability and purchasing behaviors during the COVID-19 pandemic: The importance of trust and perceived stress. *Nutrients, 12*(9), 2852.

Jia, Y., Yan, J., Liu, T., & Huang. J., (2019). How does internal and external CSR affect employees' work engagement? Exploring multiple mediation mechanisms and boundary conditions. *International Journal of Environmental Research and Public Health, 16*(14), 1–17.

Kim, P., (1992). Does advertising work: A review of the evidence. *Journal of Consumer Marketing, 9,* 5–21.

Köksal, M. H., & Özgul. E., (2007). The relationship between marketing strategies and performance in an economic crisis. *Marketing Intelligence & Planning, 25*(4), 326–342.

Laato, S., Islam, A. N., Farooq, A., & Dhir. A., (2020). Unusual purchasing behavior during the early stages of the COVID-19 pandemic: The stimulus-organism-response approach. *Journal of Retailing and Consumer Services, 57*(102224), 1–12.

Lee, J. Y., Fang, E., Kim, J. J., Li, X., & Palmatier. R. W., (2018). The effect of online shopping platform strategies on search, display, and membership revenues. *Journal of Retailing, 94*(3), 247–264.

Lee, J., Snidermann, B., Marquard, B., Galletto, N., Geeanpersadh, P., & Cherny. M., (2020). *Embedding Trust into COVID-19 Recovery.* Retrieved from: https://www2.

deloitte.com/us/en/insights/economy/covid-19/building-trust-during-covid-19-recovery.html (accessed on 18 February 2022).
Lee, M. K., & Turban. E., (2001). A trust model for consumer internet shopping. *International Journal of Electronic Commerce, 6*(1), 75–91.
Manuel, T., & Herron. T. L., (2020). An ethical perspective of business CSR and the COVID-19 pandemic. *Society and Business Review*.
Mazzoleni, S., Turchetti, G., & Ambrosino. N., (2020). The COVID-19 outbreak: From "black swan" to global challenges and opportunities. *Pulmonology, 26*(3), 117, 118.
Mogaji, E., (2020). Financial vulnerability during a pandemic: Insights for coronavirus disease (COVID-19). *Research Agenda Working Papers, 2020*(5), 57–63.
Mohan, V., Berg, C., & Poblet. M., (2020). Panic, information and quantity assurance in a pandemic. *Information and Quantity Assurance in a Pandemic*.
Moorman, C., Deshpande, R., & Zaltman. G., (1993). Factors affecting trust in market research relationships. *Journal of Marketing, 57*(1), 81–101.
Moorman, C., Zaltman, G., & Deshpande. R., (1992). Relationships between providers and users of market research: The dynamics of trust within and between organizations. *Journal of Marketing Research, 29*(3), 314–328.
Morton, F., (2018). The use of written descriptions and 2D images as cues for tactile information in online shopping. In De, S., Bhattacharyya, S., & Dutta. P., (eds.), *Intelligent Multidimensional Data and Image Processing* (pp. 18–37). IGI Global.
OECD, (2020). *Building Back Better: A Sustainable, Resilient Recovery After COVID-19*. Retrieved from: https://read.oecd-ilibrary.org/view/?ref=133_133639-s08q2ridhf&title=Building-back-better-_A-sustainable-resilient-recovery-after-Covid-19 (accessed on 18 February 2022).
Piercy, N. F., Cravens, D. W., & Lane. N., (2010). Thinking strategically about pricing decisions. *Journal of Business Strategy, 31*(5), 38–48.
Security, (2020). *Cybersecurity Experts Come Together to Fight Coronavirus-Related Cyberattacks*. Retrieved from: https://www.securitymagazine.com/articles/92004-cybersecurity-experts-come-together-to-fight-coronavirus-related-cyberatacks (accessed on 18 February 2022).
Sharma, V., & Sonwalkar. J., (2013). Does consumer buying behavior change during an economic crisis? *International Journal of Economics & Business Administration, 1*(2), 33–48.
Shin, H., & Kang. J., (2020). Reducing perceived health risk to attract hotel customers in the COVID-19 pandemic era: Focused on technology innovation for social distancing and cleanliness. *International Journal of Hospitality Management, 91*, 102664.
Sirdeshmukh, D., Singh, J., & Sabol. B., (2002). Consumer trust, value, and loyalty in relational exchanges. *Journal of Marketing, 66*(1), 15–37.
Song, W., Li, W., & Geng. S., (2020). Effect of online product reviews on third parties' selling on retail platforms. *Electronic Commerce Research Applications, 39*(100900), 1–12.
Taleb, N., (2008). *The Black Swan: The Impact of the Highly Improbable*. London: Penguin.
Tran, L. T. T., (2020). Managing the effectiveness of e-commerce platforms in a pandemic. *Journal of Retailing and Consumer Services, 58*, 102287.
Urban, G. L., Sultan, F., & Qualls. W. J., (2000). Placing trust at the center of your internet strategy. *Sloan Management Review, 42(Fall)*, 39–49.

Wang, Y., Xu, R., Schwartz, M., Ghosh, D., & Chen. X., (2020). COVID-19 and retail grocery management: Insights from a broad-based consumer survey. *IEEE Engineering Management Review, 48*(3), 202–211.

Werner, L. R., (1991). Marketing strategies for the recession. *Management Review, 80*(8), 29–30.

Zand, D. E., (1972). Trust and managerial problem solving. *Administrative Science Quarterly, 17*(2), 229–239.

Zwanka, R. J., & Buff. C., (2020). COVID-19 generation: A conceptual framework of the consumer behavioral shifts to be caused by the COVID-19 pandemic. *Journal of International Consumer Marketing,* 1–10.

CHAPTER 7

How Restrictive Conditions Require Marketing Decisions Based on Analytics and Predictive Modeling: A Description of Two Cases

JUAN IGNACIO GONZALEZ ESPINOSA

Engineering Management Department at University of Monterey, Mexico, E-mail: jigonzale040@gmail.com

ABSTRACT

As effects of pandemic COVID-19 emerge and become present in the retail, branding, and general marketing arena nowadays, it is evident and imperative to look for new veins of effective decision taking over the pitfalls of Pandemics. While customer purchasing and consumption habits change (in-home behavior), firms confront a shortage of physical attendants on "brick and mortar" stores and significantly increase digital commercial transactions. Big Data may be part of the solution for facing this new business landscape. As data is more and more available about consumer behavior, branding, and distribution channels, descriptive, predictive, and prescriptive analytics may support and improve decision-making in these areas mentioned above. This chapter exemplifies this using two real-world cases: one based on segmenting industrial clients, operating in a web-based sales platform, and the second, explaining a model for preventing stock-out from shelves. As cases demonstrate, the use of data collected by firms may represent a valid, helpful platform for new insights in decision making to cope effectively with the pandemic COVID-19 environment.

Marketing by Contingency in the Time of COVID-19: Overcoming Business Crises and Meeting Marketing Challenges. Alicia De La Peña, PhD & Juan Bernardo Amezcua Nunez, PhD (Eds.)

7.1 INTRODUCTION

As recent times establish challenging, different, and in some sense, chaotic conditions in the local, national, and international scenario (e.g., COVID-19), the social, cultural, and economic environments have been touched by distinctive characteristics that challenge the way we currently understand, manage, and decide.

Significantly, decision-making in business in general and particular areas has been affected in a relevant way. As our new reality becomes part of the concrete establishment and current decision making, it is evident that the traditional processes for planning and execution may not be sufficient and convenient, neither profitable.

A proof of this is the emerging analysis of the consequences and effects of pandemic COVID-19. Jason Goldberg, in his presaging article published in FORBES in March 2020 (Goldberg, 2020), mentions three main forces landscaping brands and retailing in the COVID-19 era: 1) adoption of short-term consumer behaviors, 2) new economic reality, and 3) a brand-new competitive and partnership ecosystem.

About the short-term consumer behaviors, Goldberg's estimates can be permanent: it can be found a profound increasing trend of digital commerce use (210%, considering the period of March 12th to March 15th, 2020, comparing with the same period of the preceding year [Goldberg, 2020]). This effect provoked the closure of hundreds of "brick and mortar" retail's physical locations and stores, as consumers are more concerned about the "germaphobia" and the "in-homing" behavior of purchasing and consumption is here to stay (Goldberg, 2020, pp. 4, 5).

In terms of the new competitive and partnership environment, Goldberg explains that physical (foot) traffic fell off by 58.4% just by the third week of March 2020 in US stores and retailers (The Wall Street Journal, 2020), and 78.9% for apparel. The consequences of this foot-client shortage provoked the reconversion of business models (turned to digital commerce) and the slimming down of personnel structures. Digital disruption is a reality.

But there is another side of the story. Some US retailers revealed a little or no impact from COVID-19 Contingency. As retail and food services sales in general decreased 7.7% by February to April 2020 in front of the same period in 2019, the ones from grocery stores and non-store retailers (e-commerce providers) increased by 16% and 14.8% (respectively) in the same period (OECD, 2020). A corresponding explosion of online-purchasing apps (such as Walmart Grocery, Instacart, and Shipt) rocketed to unimaginable download rates by March 15th (Blacker, 2020). Bradlow, Gangwar, Kopalle, and Voleti (2017) pointed out that Walmart collects an insane amount of information about clients every hour: 2.5 petabytes (1 petabyte = 1,000,000 gigabytes). Also, the IT consulting Gartner estimates that there will be 20 million devices connected to the "internet of things" by 2020, interacting with people, computers, and other personal devices (Hung, 2017). As Shankar (2019) stated, it may not be reckless to say: "The future will be dominated by AI-assisted customer behavior and AI-based managerial decisions" (p. 40). How do data and its availability play on this new horizon?

One particular choice to opt for under these circumstances is to look at the notion of Big Data and its availability in quantity and quality. As more and more data is available, it is possible to establish focused research and use data for developing more intelligent, directed, and real-time decisions for specific and focused goals in decision making (Jeble, Kumari, and Patil, 2018).

The role of prediction has been relevant under these new conditions for decisions and the environment we face (McAfee, Brynjolfsson, Davenport, Patil, and Barton, 2012). Predictive capabilities in the firm evolve as an effective way for constructing scenarios and establishing methods for action as reality started to emerge in the sea of complexity (Pagallo, 2018), and also necessary to recognize that these predictive algorithms require the human conjecture and sense-making, as stated by Pridmore and Hämäläinen (2017).

Regardless of the relevance of the RBV (resources based view) of the firm in our discussion context (Adrian, Abdullah, Atan, and Jusoh, 2018), the relevance of landing the reality of the descriptive, predictive, and prescriptive analytics models in today's business decision making platforms is mandatory. Fosso-Wamba, Akter, Edwards, Chopin, and Gnanzou (2015), described how corporate decision making evolves for addressing environmental changes more rapidly and proactively, as decision takers possess more significant knowledge of relationships among relevant variables and factors involved.

Maybe, uncertainty, and lack of information were the most challenging factors for producing convenient and approachable decisions before the '90s. Big Data improves decision-making, reducing those factors, as it is possible to transform data into usable information (Tihanyi, Graffin, and George, 2014). This is a fact on our new reality, settled by COVID-19.

In terms of discussing specifically the Marketing discipline, the book *Creating value with big data analytics: Making smarter marketing decisions*, by Verhoef, Kooge, and Walk (2016), specifies the benefits of marketing decisions using BD in terms of various dimensions: market, brand, and consumer levels, and the value produced in terms of value to the firm and value to the consumer. This exciting combination of advantaging results, according to the authors, emerges from the integration of fragmented, unrelated, varied data, synchronization of analytical perspectives on market segments, history available, and identifying and linking value produced both for the firm and for the customer (see chapter: *Data Integration*).

In another perspective, but the same line, Krush, Agnihotri, and Trainor (2016) agree in considering marketing measurement and analytics in a "fierce global competition and rapid technological change" (p. 2078), reaching annual calculated global expenditures for USD 24 billion by 2018, in software systems (Lazich, Burton, and Virgil, 2016).

In their publication, Krush, Agnihotri, and Trainor (2016) determined a model for explaining the market performance of firms using the speed implementation of marketing strategy, market information, and management capability of the firm; all this framework contained in the use of Marketing dashboards and the presence of technological uncertainty.

Ducange, Pecori, and Mezzina (2018) punctuate that the transformation of the hefty volume of information into accurate, relevant information for decision-making in marketing, stands for an essential ability of firms. The capability to implement analysis and correlation of that information is a requisite for creating firm value, growth, productivity, and innovation (p. 325). In the same line, they mention some fundamental roles and important establishments for doing this: machine learning techniques, computational intelligence, and other more qualitative approaches (as text mining, sentiment analysis, social sensing) all are supporting factors for deploying helpful analysis in the marketing decision arena. Moreno et al. (2016); and Jha et al. (2016) set approaches that confirm this statement.

Ducange, Pecori, and Mezzina (2018) have developed an exciting literature review on marketing strategies based on big data and analytics, on these main categories:

- Social big data as an integration of traditional marketing strategy;
- Fostering commercial and promotional activities;

- Analysis of perception and reputation; and
- Management of customer relationships.

Regardless of analyzing the former compilation of Ducange, Pecori, and Mezzina (2018), and other relevant work in the literature, it is convenient to assess and comprehend specific cases that demonstrate in detail how the deployment of analytics algorithms in Marketing supported complex, relevant decisions, bonded with meaningful and pertinent market activities of two Mexican firms, operating in the Northern zone of the country.

The next section of this chapter is devoted to the development of these stories. The first one is linked with discovering client behavior on web-based platforms focused on all the CRM processes, specifically, the one related to the purchase, for a metal-mechanic firm. The second case is associated with discovering factors related to the in-store shelves stock-outs for a retailing supermarket chain.

All cases were developed by a research team guided by a leader, an academic with a PhD and specialized in analytics. The three points were created using the CRISP-DM© methodology (IBM, 2012). The CRISP-DM© method comprehends the following stages that were developed by the three research teams:

1. **Business Understanding:** This initial phase focuses on understanding the project objectives and requirements from a business perspective and how to convert this knowledge into a data mining problem definition and a preliminary plan designed to achieve the goals.
2. **Data Understanding:** The data understanding phase begins with initial data collection. It continues with activities that allow the research team to become familiar with the data, identify data quality issues, discover early insights from the data, and detect interesting subsets to form hypotheses about confidential information.
3. **Data Preparation:** The data preparation phase covers all activities necessary to build the final data set (data incorporated into the modeling tool(s) from the initial raw data). Data preparation tasks are likely to be performed multiple times and not in any prescribed order. Duties include selecting tables, records, attributes, and transformation and cleaning of data for modeling tools.
4. **Modeling:** Various modeling techniques are selected and applied in this phase, and their parameters are calibrated for optimal

values. Typically, there are several techniques for the same type of data mining problem. Some methods have specific requirements on the shape of the data. Therefore, it is often necessary to go back to the data preparation phase.

5. **Evaluation:** At this stage of the project, the research team has built a model (or models) that appears to be of high quality from a data analysis perspective. Before proceeding with the final deployment of the model, it is essential to thoroughly evaluate and review the steps executed to create it to ensure that the model adequately achieves the business objectives. A key aim is to determine if any significant business issues have not been considered with the importance required. At the end of this phase, a decision must be made on using the data mining results.
6. **Deployment:** Model creation is generally not the end of the project, even if the purpose of the model is to increase knowledge. The knowledge gained from the data must be organized and presented so that the client can use it. This often involves applying "live" models within an organization's decision-making processes, for example, real-time personalization of web pages or repeated scoring of marketing databases. Depending on the requirements, the implementation phase can be as simple as generating a report or as complex as implementing repeatable data mining across the enterprise. In many cases, the customer is not the data analyst who carries out the implementation steps. However, even if the analyst will carry out the implementation effort, the customer must understand what must take actions to use the models created.

7.2 CASE 1: CLUSTERING FOR WEB-BASED CLIENTS

7.2.1 CONTEXT AND METHOD

Currently, as firms develop operations and perform in complex environments, valid concepts, like Industry 4.0, emerge as primary platforms for acquiring excellence in performance. As a result of automation of main business processes, data availability promotes all sorts of data analytics, automation, and visualization projects, pursuing the improvement of operations and decision-making.

This is the case of a Norther-Mexican, metal mechanic firm concerned about the competitive edge prevailing in the steel industry. Operating in Mexico for many years ago, this firm produces and sells a variety of finished, semi-finished, and OEM (Original Equipment Manufacturer) products countrywide. In the last 5 years, the firm has embarked on an ambitious program of automation of operative and commercial business processes, setting a goal of creating a data-driven firm for the next years. Remarkably, the firm has implemented a web-based platform for serving customer transactions digitally, from doing product searches to deploying the final sales of products.

As the platform has evolved, and more clients use its services instead of the traditional attention desk and sales representatives, a deep knowledge of customer behavior were needed, as long as the understanding about what clients do inside the service became a black box: some clients completed the purchase all the way, but others finished connection in the very early stages of the process. What is more: some other clients did not show interest in using the platform, although deployed an immense and focused training effort.

A key question emerged: how to discover the hidden behavior of firms inside the platform and get acquainted with factors arousing this variety of results?

Considering an investigative effort developed by Martínez Aldape, Espinosa Ramírez, Cabello Hernández and Quintanar Villarreal (2019), a phase of deep understanding of consumer behavior started. After developing various clustering unsuccessful attempts, the research team decided to implement a deep analysis of the "natural" stages of the purchasing process inside the web platform. The team discovered these four stages along the process:

1. **"Tree" Phase:** Here clients check out the stock for the existences of the desired product according to the nearest or desired distribution center from shipment.
2. **Selection Phase:** In this stage, clients select the quantity of the product to be purchased and establish its specifications.
3. **"Shopping Cart" Phase:** At this point, clients quote the selected products, review specifications and prices.
4. **Confirmation Phase:** Finally, the customer chooses the type of shipment and completes the purchase transaction. This is the "target" phase for the firm.

As these stages delineate the whole process, the research team identified the client base into these four categories, as they accomplished one of the four stages explained above. Every client was categorized according to a dominant behavior they showed in 8 months of analysis (January–August, 2019). A specific example will clarify the categorization process: as client Alpha, in terms of the 100% of its connections during the time analyzed, 90% of the times completed the purchase. So this client was categorized in the Confirmation phase. Criteria used then relied on the most frequent behavior of clients. Then, Table 7.1 shows the total base of clients classified and named as shows.

TABLE 7.1 Categorization of Clients (Name of the Cluster) by the Phase They Were Assigned

Phase	Cluster
"Tree"	Navigator
Selection	Selector
"Shopping cart"	Estimator
Confirmation	Shopper

Source: Own elaboration.

By the time of analysis was effectuated, 21% of clients were navigators, 24% were selectors, 10% were estimators and 45% were shoppers. A provocative question emerged under these results: regardless of shoppers' current figure, how could the firm raise this percentage?

As a part of the corporate marketing strategy of the firm, indeed, there were a set of promotional incentives planned to direct to the client's firm base. Although, this new evidence requires rethinking the course of action. In this way, contingent factors emanated from the development of current client behavior offer new insights for action.

The research team decided to move to the next stage. Develop predictive models for identifying relevant factors that may provoke navigators, selectors, and estimators migrating to shoppers.

Regardless of the potential economic benefits, acquiring a more extensive base of shoppers represents an essential goal for positioning and loyalty for the firm. Decided to implement predictive modeling for discovering the triggering factors for promoting Estimators to Shoppers. A migrating base of 10% of clients is a worthwhile gathering.

The first step was to discover relevant variables for including in the predictive modeling. Table 7.2 shows the list of relevant variables considered and obtained from the available databases in the firm.

TABLE 7.2 List of Relevant Variables and Description for Implementing Predictive Modeling

Variable	Description
GDP per capita by state	A measure of economic dynamics assigned by the state in which the client operates primarily.
Number of client sessions	The number of sessions performed in the period analyzed by the client.
Number of purchases of total sessions performed	A ratio of purchasing transactions over complete sessions performed by the client.
Average clicks by session	The average number of clicks performed by the client, by session.
Number of tons quoted	Sum of tons quoted, considered in the time analyzed.
Number of tons purchased.	Sum of tons purchased, considered in the time analyzed.
(Continued) Number of Tons quoted, normalized.	Sum of tons quoted, considered in the time analyzed. Normalized.
Number of tons purchased, normalized.	Sum of tons purchased, considered in the time analyzed. Normalized.
Negotiation	The request of a reconsideration of the total price of a quoted order. Dichotomous variable: 1 = client asked for a price review, 0 = no request solicited
Order allocating channel	A dichotomous variable: whether the client places his/her order using the web platform (0) or by the corresponding sales representative (1)
Shipping method	Dichotomous: whether client receives his/her product shipped by the firm (0). or picks up at the nearest distribution center (1)

Source: Own elaboration.

7.2.2 RESULTS AND IMPLICATIONS

The research team decided to implement a predictive model based on a Logistic Regression. According to Hosmer, Lemeshow, and Sturdivant (2013), this predictive model is parsimonious and easy to understand. This non-linear regressive supervised model emerges as the trained algorithm for estimating a dependent variable with a range of 0 to 1, allowing interpreting these values as probabilities (Wright, 1995). A threshold value of 0.75 was established for the probability of pushing a client and considering it as a future Shopper. In this way, as every client present in the used databased was assigned a value of 0 (Estimator) and 1 (for Shopper), results of the model allowed the research team to calculate a probability for estimating a future behavior of Estimator or Shopper, under specific values of independent variables. Table 7.3 shows the results of the final logistic regression, developed in SPSS (Statistical Package for Social Sciences) ver. 22, for the 8 months of analysis.

TABLE 7.3 Final Results for Logistic Regression: Predicting Shopper's Behavior from Estimators and Shoppers

Variables		S.E.	Wald	Df	Sig.	Exp (β)
Number of tons purchased, normalized.	21.77	2.12	105.55	1	0	2868852328
Negotiation	0.786	0.181	18.79	1	0	2.195
Shipping method	–0.678	0.266	6.52	1	0.011	0.508
Model constant	8.79	0.778	127.516	1	0	6,570.505

Significance (alfa) = 10%

Source: Own elaboration.

Going beyond the math, the best model throw light on the number of tons purchased (as logic may suggest), but other interesting variables emerge: negotiation and the shipping method. In other words, essential factors promoting Estimators for becoming Shoppers in the future are the purchasing results, requesting negotiation, and receiving the product shipped by the firm (note the negative sign of the Beta of variable *Shipping Method*).

The model showed a relatively good performance identifying Estimators and Shoppers. Table 7.4 reflects the confusion matrix (table of correct classifications) of the model presented above.

TABLE 7.4 Confusion Matrix (Table of Correct Classifications) for the Final Regression Model

		Predicted Category		% Correct Predictions
	Observed	**Estimator**	**Shopper**	
Category	Estimator	204	97	67.8%
	Shopper	80	943	92.2%
	% General Correct Predictions			86.6%

Source: Own elaboration.

As can be inferred from Table 7.4, the model has an overall accuracy of 86.6%. Another relevant measure of adjustment is Recall = 71.83%, Precision = 67.8% and F-measure = 69.7%. Due to the imbalanced sample, the model is more suitable for predicting Shoppers than Estimators. This is certainly a drawback for the current model and future analytics efforts need to be considered at this point.

Some valuable learnings that can be inferred from the model developed are:

- Under the pressure of the current economic environment, the model suggests that the concept of a preferential price to clients need to be adhered to the web system to enhance its effect and power, as clients highly value a price reconsideration and obviously, positive results for completing the purchase in the platform. Additional "external" price incentives may not have power outside the web platform.
- On the other side, shipment decisions may be pondered on the side of the client. As the model suggests, clients picking up their products at the nearest distribution center may advance to become eight shoppers.
- As the two prior numbered bullets suggest, the management and allocation of resources and decisions about price and distribution can follow the model's path developed, as it is possible to measure their effectiveness. In this way, as the firm uses available data of clients, it may derive (and should do) insights from analytics' predictive models and concentrate on the better decision for resources, talent, and strategic efforts allocation in marketing.

7.3 CASE 2: PREVENTING PRODUCT IN-SHELF STOCKOUTS

7.3.1 CONTEXT AND METHOD

The retailing industry across the world is evolving. Reducing margins, the meteoric rise of competency, and online sales gaining space on consumer preferences. Under these specific circumstances, in-store, and shelf space value has become the best asset for retailers.

In this sense, optimized use of these valuable spaces is a priority for physical stores. According to Frontoni, Marinelli, Rosetti, and Zingaretti (2017), it is essential to maintain the right assortment of products to acquire satisfied customers and enhance loyalty from a more extensive client base. Incentives inside the store are also significant, but the absence of products on the shelf is, maybe, the worst scenario for retailers. Missing products may cause, in terms of the discussion of Frontoni et al. (2017, p. 32), some of these outputs:

- Getting a substitute in the store;
- Postpone the desire;
- Avoid purchasing;
- Purchasing in another location (the worst-case scenario).

Despite all relevant implications of product allocation in the store, such as in-store logistics, avoiding misplacement, and faulty transactions (Chuang, 2018), stock-outs in shelf space become one most critical factor for a base of loyal clients and profitable operations.

This idea serves as the foundation of this case. A retailing chain operating in Mexico challenged our research team for modeling the factors causing shelf stock-outs. The focus of this modeling exercise was the popular, middle-sized store, focused on low prices, with limited product lines and primarily focused on food. Data used in this chapter were collected by Sepúlveda, Tamez, and Flores (2018).

The retailer faces a fierce competency in this segment. And with reduced margins, maintaining a convenient level of inventory becomes critical. The retailer wanted to know, before making a costly decision in terms of in-store inventory and providing time, critical factors influencing the shelf out of stock. Especially, if the current level of shelf out-of-stock for the segment analyzed accounted for 1.6% of total sales, whereas the firm goal was 0.5%.

As the first step, the research team implemented a literature review concerning the stock-out phenomenon. As confronting theory and reality, the team reached archival data from the retailer and obtained available information. A database was constructed, considering a daily record of available data. The resulting set of acquired variables is shown in Table 7.5.

TABLE 7.5 List of Relevant Variables for Modeling the Stock Out in the Research Context

Variable	Description
Quality and freshness of products	Internal evaluation
Availability of free check-outs	
Help on the purchasing process	
Order and accommodation of products	
Help offered by the voluntary packer	
Sensorial perception of the store	
Age of client	
Amount of time for paying products	
The external appearance of the store	
Kindness and training of store employees	
Quickness of cashier	
The general perception of the store	
Experience of the purchasing process	
Perception of the value obtained.	
Availability of products	
Perception of promotional activities in store	
Evaluation of product's price	Internal evaluation
Evaluation of competitors' product price	External agency
Age of store	Internal data
The average temperature of the day	Weather channel.com
Rainfall in cubic centimeters of day	Weather channel.com
Average daily sales of products observed	Internal evaluation
Average daily sales of complementary products	Internal evaluation
Shelf capacity	Internal evaluation
Product front in shelf (number of products in shelf facing client)	Internal evaluation
Number of people for fulfilling shelves	Internal evaluation
Fulfilling schedule	Internal evaluation
Supply fulfilling schedule	Internal evaluation
Days of inventory in store	Internal evaluation

Source: Own elaboration.

The research team faced the challenge of selecting the product sub-families to model, as various product groups faced stock-outs. The analysis required tracking data for 52 weeks in 2018. After developing an analysis of importance, the team decided to focus on two sub-families: dairy and beverages and snacks, and specifically, construct models for the following categories:

- Eggs (white, medium);
- Drinkable yogurts;
- Liquid cream;
- Margarine;
- Meat;
- Solid produce;
- Juices;
- Cola (generic Coke and Coke) drink, regular;
- Cola (generic Coke and Coke) drink, light;
- Cola flavored, regular;
- Water, flavored;
- Snacks;
- Popcorn;
- Children drinkables.

Modeling work consisted of the development of 14 applications based on the categories listed. Models developed, based on logistic regression, varied from 50% to 97% for general accuracy. The research team decided to use the logistic model for estimating the probability of future shelf stock-out for the categories listed. A variety of insights emanated from models developed. For the sake of brevity, only two models and results will be shown: children drinkables and light cola drinks.

7.3.2 RESULTS AND IMPLICATIONS

7.3.2.1 MODEL 1: BEVERAGES FOR CHILDREN

This category comprehends various beverages (like yogurt and flavored, enriched liquids) aimed at children (0 to 12 years old). Table 7.6 shows the final logistic model for the category, specifying the independent variables affecting the occurrence of shelf stock-outs.

TABLE 7.6 Final Results for Logistic Regression: Predicting Shelf Stock Out for Category Children Drinkables

Variable		S.E.	Z-Value	Sig.
Model constant	11043.7	–1271.706	–8.684	0.000
Promotional activities of product	932.690	99.781	9.347	0.000
The external appearance of the store	–454.322	52.008	–8.736	0.000
Promotions and specials	–548.824	63.067	–8.702	0.000
Perception of the value obtained	397.631	45.851	8.672	0.000
Quickness of cashier	–889.429	102.424	–8.684	0.000
Average daily sales of products observed	–19.953	2.314	–8.621	0.000
Quality and freshness of products	713.063	86.145	8.277	0.000
Age of client	27.595	3.301	8.361	0.000
Help on the purchasing process	–621.800	76.618	–8.116	0.000
Availability of free check-outs	197.437	24.937	7.918	0.000
Availability of products	–229.830	30.706	–7.485	0.000
Product front in shelf (number of products in shelf facing client)	50.835	6.851	7.420	0.000
Shelf capacity	1.545	0.217	7.110	0.000
Order and accommodation of products	–72.908	10.314	–7.069	0.000
Help offered by the voluntary packer	–383.640	60.083	–6.385	0.000
Shelf capacity	–18.093	2.879	–6.284	0.000
Average daily sales of complementary products (Hygiene and personal care)	0.036	0.011	3.291	0.001
Average daily sales of complementary products (no-food groceries)	–0.006	0.002	–3.084	0.002
Days of inventory in store	0.343	0.159	2.149	0.032

1 = Shelf stock-out; 0 = No stock-out; Significance (alfa) = 5%

Source: Own elaboration.

It is interesting how to interpret the negative coefficients of variables in Table 7.6. These, in a practical interpretation, may prevent a stock-out, whereas those variables with positive betas would raise a stock-out probability. In this case, a stock-out implies good news, as these factors would foment products disappearing from the shelf-successful sales, in other words-, but an excellent opportunity for operative planning in preventing a possible product blackout. Perception of product freshness, promotional activities in store supporting the category, and guaranteeing a fast check-out will enhance the sales of this category. Evaluating the model presented, as Table 7.7 presents, demonstrates good prediction of shelf stock-outs.

TABLE 7.7 Confusion Matrix (Table of Correct Classifications) for the Final Model of Children Drinkables

		Predicted Category		**% Correct Predictions**
	Observed	**Stock-Out**	**No Stock-Out**	
Category	Stock-Out	95	3	96.94%
	No Stock-Out	2	262	99.24%
	% General Correct Predictions			98.61%

Source: Own elaboration.

As the accuracy shows 98.61% of correct predictions, the model becomes reliable and an excellent source of decision-making insights.

7.3.2.2 MODEL 2: LIGHT COLA (GENERIC BRANDS AND COKE)

Products in this category are essential for the firm. It must be emphasized that Mexico is the largest consumer of Coca-Cola and comparable products worldwide (Fortune, 2019). Therefore, the relevance for the segment studied is evident. Additionally, selling the leader brands (Coca-Cola and Pepsi also commercialize imported brands like Big Cola and private ones. This category is susceptible to promotional activities in store, as well as specific events, like holidays and celebrations.

In Table 7.8, the final logistic model is shown, describing the relevant variables affecting the shelf stock-out for this category.

TABLE 7.8 Final Results for Logistic Regression: Predicting Shelf Stock-Out for Category Cola (Coke and Generic Coke)

Variable		S.E.	Z-Value	Sig.
Model constant	19.275	2.209	–8.726	0.000
Shelf capacity	–0.180	0.032	–5.548	0.000
Days of inventory in store	–0.195	0.049	–4.011	0.000
Experience of the purchasing process	–3.324	0.845	–3.934	0.000
Kindness and training of store employees	–0.522	0.140	–3.729	0.000
Evaluation of product's price	2.067	0.584	3.542	0.000
Availability of products	2.670	0.858	3.113	0.002
Average daily sales of complementary products (Health category)	–0.003	0.001	–3.086	0.002
Age of client	–0.255	0.084	–3.042	0.002
Availability of free check-outs	1.575	0.643	2.448	0.014
Average daily sales of complementary products (no-food groceries)	0.000	0.000	1.783	0.075

1 = Shelf stock-out; 0 = No stock-out; Significance (alfa) = 5%.

Source: Own elaboration.

In contrast to the former category developed (children drinkables), light Coke and generic Coke beverages are linked with the product's availability, fast service, and convenient price. Available space and days of inventory may prevent stock-outs. It seems that juvenile customers will take the product from shelves. A relevant fact must be emphasized: it appears that the category is complimentary for the Health product family (see the negative coefficient beta for the variable) (Table 7.9).

TABLE 7.9 Confusion Matrix (Table of Correct Classifications) for the Final Model for the Cola Category (Coke and Generic Coke)

		Predicted Category		% Correct Predictions
	Observed	**Stock-Out**	**No Stock-Out**	
Category	Stock-Out	36	15	82.32%
	No Stock-Out	55	256	70.59%
	% General Correct Predictions			80.66%

As the model predicts 80.66% in a global performance, it is pretty suitable for interpreting.

7.4 CONCLUSION

By the cases described above, it is evident how the use analytics models (fed with available data) provides meaningful insights about the surrounding conditions of the firm's products and services. Without any doubt, these models are valuable sources of better decisions focused on gaining a competitive advantage. While the first case emphasizes a segmentation initiative based on online customer behavior, the second one offers a practical approach for preventing stock-outs on the store shelf.

Valuable insights emerge from a specific, "pure" segmentation effort produced entirely by the online purchasing behavior, especially from a B2B market, as more and more corporations will evolve to electronic business models for interacting with clients. The second context described a set of specific models for estimating shelf stock-out in store. As described in the introductory section of this chapter, patterns of purchasing behavior and consumption have changed, and the rise of grocery delivery services (as Walmart Grocery, Instacart, Ship, or Amazon Fresh) are reshaping the retail activity. Before the Pandemic, final consumers visited the sales floors of department stores and supermarkets, looking for and feeling an experience while getting groceries or merchandise. Factors as "germaphobia" and forced home reclusion have boosted the demand for grocery delivery services. Agents of these services are far from looking for a shopping experience (e.g., perceiving different scents, textures or lighting in the store); they are focused on fast and effective service, product stocks and convenient locations for accessing, as they may complete more deliveries in one day.

Appreciating all this evidence, big data and data analytics (descriptive, predictive, and prescriptive models) could be considered as an alternative and specific strategic decision platform for decision making, especially when traditional business models may be no longer valid in front of COVID-19 Pandemic and its new establishment. Data can be, in this case, effective, and appropriate for businesses suffering the effect of the COVID-19 virus.

KEYWORDS

- **analytics models**
- **confusion matrix**

- **generic coke and coke**
- **health product family**
- **shelf stock-outs**

REFERENCES

Adrian, C., Abdullah, R., Atan, R., & Jusoh. Y. Y., (2018). Conceptual model development of big data analytics implementation assessment effect on decision-making. *International Journal of Interactive Multimedia and Artificial Intelligence, 5*(1), 101–106.

Ashander, L., Kliestikova, J., Durana, P., & Vrbka. J., (2019). The decision-making logic of big data algorithmic analytics. *Contemporary Readings in Law and Social Justice, 11*(1), 57–62.

Blacker, A., (2020). *Instacart and Grocery Delivery Apps Set Consecutive Days of Record Downloads.* Apptopia. https://blog.apptopia.com/instacart-and-grocery-delivery-apps-set-consecutive-days-of-record-downloads (accessed on 18 February 2022).

Bradlow, E. T., Gangwar, M., Kopalle, P., & Voleti. S., (2017). The role of big data and predictive analytics in retailing. *Journal of Retailing, 93*(1), 79–95.

Chuang, H. H. C., (2018). Fixing shelf out-of-stock with signals in point-of-sale data. *European Journal of Operational Research, 270(3)*, 862–872.

Ducange, P., Pecori, R., & Mezzina. P., (2018). A glimpse on big data analytics in the framework of marketing strategies. *Soft Computing, 22*(1), 325–342.

Fortune, (2019). *México es el Mayor Consumidor de Refresco en el Mundo (Mexico is the World's Largest Soft Drink Consumer).* https://www.fortuneenespanol.com/noticias/mexico-encabeza-lista-mundial-consumidores-refresco (accessed on 18 February 2022).

Fosso, W. S., Akter, S., Edwards, A., Chopin, G., & Gnanzou. D., (2015). How "big data" can make big impact: Findings from a systematic review and a longitudinal case study. *International Journal of Production Economics, 165*, 234–246.

Frontoni, E., Marinelli, F., Rosetti, R., & Zingaretti. P., (2017). Shelf space re-allocation for out-of-stock reduction. *Computers & Industrial Engineering, 106*, 32–40.

Goldberg, J., (2020). *The Impact of Covid-19 on U.S. Brands and Retailers.* FORBES. https://www.forbes.com/sites/jasongoldberg/2020/03/29/the-impact-of-covid-19-on-us-brands-and-retailers/?sh=2543a2311452 (accessed on 18 February 2022).

Hosmer, Jr. D. W., Lemeshow, S., & Sturdivant. R. X., (2013). *Applied Logistic Regression.* New York: John Wiley & Sons.

Hung, M., (2019). *Leading the IoT. Gartner Insights on How to Lead in a Connected World.* Gartner Inc. https://www.gartner.com/imagesrv/books/iot/iotEbook_digital.pdf (accessed on 18 February 2022).

IBM, (2021). *CRISP-DM© Help Overview.* IBM Knowledge Center. https://www.ibm.com/support/knowledgecenter/en/SS3RA7_sub/modeler_crispdm_ddita/clementine/crisp_help/crisp_overview.html (accessed on 18 February 2022).

Jeble, S., Kumari, S., & Patil. Y., (2018). Role of big data in decision making. *Operations and Supply Chain Management: An International Journal, 11*(1), 36–44.

Jha, A., Dave, M., & Madan. S., (2016). A review on the study and analysis of big data using data mining techniques. *International Journal of Latest Trends in Engineering and Technology, 6*(*3*), 94–102.

Krush, M. T., Agnihotri, R., & Trainor. K. J., (2016). A contingency model of marketing dashboards and their influence on marketing strategy implementation speed and market information management capability. *European Journal of Marketing, 50*(*12*), 2077–2120.

Lazich, R. S., Burton, I., & Virgil. L., (2016). *Global Sales of Marketing Software, 2014–2018*. Gale Group, Michigan.

Martínez, A. A. L., Espinoza, R. J. G., Cabello, H. T. V., & Quintanar. V. J. L., (2019). *Development of Analytics Models to Understand Customer Behavior in a Mexican Metal-Mechanic Firm [Unpublished Undergraduate Dissertation].* Universidad de Monterrey (UDEM).

McAfee, A., Brynjolfsson, E., Davenport, T. H., Patil, D. J., & Barton. D., (2012). Big data. The management revolution. *Harvard Business Review, 90*(*10*), 61–67.

Moreno, M. V., Dufour, L., Skarmeta, A. F., Jara, A. J., Genoud, D., Ladevie, B., & Bezian. J. J., (2016) Big data: The key to energy efficiency in smart buildings. *Soft Comput., 20*(*5*), 1749–1762.

OECD (2020). E-commerce in the time of COVID-19. *OECD Policy Responses to Coronavirus (COVID-19)*, https://www.oecd.org/coronavirus/policy-responses/e-commerce-in-the-time-of-covid-19-3a2b78e8/ (accessed on 11 March 2022).

Pagallo, U., (2018). Algo-rhythms and the beat of the legal drum. *Philosophy & Technology, 31*(4), 507–524.

Pridmore, J., & Hämäläinen. L. E., (2017). Market segmentation in (In) action: Marketing and 'yet to be installed' role of big and social media data. *Historical Social Research, 42*(1), 103–122.

Sepúlveda, G. A. P., Tamez, S. C. Y., & Flores. M. M., (2018). *Implementation of Analytics Models for Predicting Shelf Stock-Outs in a Mexican Retailer*. [Unpublished undergraduate dissertation]. Universidad de Monterrey (UDEM).

The Wall Street Journal, (2020). *Coronavirus Crisis Reveals Retail Haves and Have-Nots*. https://www.wsj.com/articles/coronavirus-crisis-reveals-retail-haves-and-have-nots-11585047603 (accessed on 18 February 2022).

Tihanyi, L., Graffin, S., & George. G., (2014). Rethinking governance in management research. *Academy of Management Journal, 57*(6), 1535–1543.

Verhoef, P., Kooge, E., & Walk. N., (2016). *Creating Value with Big Data Analytics: Making Smarter Marketing Decisions*. New York: Routledge.

Wright, R. E., (1995). Logistic regression. In: Grimm, L. G., & Yarnold, P. R., (eds.), *Reading and Understanding Multivariate Statistics* (pp. 217–244). Washington, DC: American Psychological Association.

CHAPTER 8

Black Swan Instinct: Emerging Trend Models of Consumer Behavior

ANA ZAVALA-PARRALES and ANA RUIZ-MEDINA

Dean, Universidad Americana del Noreste, Coahuila, Mexico.
E-mail: azavala@uane.edu.mx

ABSTRACT

The year 2020 brought the unexpected – a decisive change in life and reality as it was known, through a worldwide emergence of a VUCA environment that became the "new normal." When faced with this type of situation, consumers were hit with new realities, and with them, a drastic change in how they were living, the decisions they made, and the patterns and trends they followed or acted upon when shopping. This chapter will focus on how COVID-19 impacted consumers by "changing the game" – how normality was perceived, what everyday life looked like, what could become a trend, and how it was suddenly disrupted by a black swan, changing every minor and major detail and aspect in people's lives in such a way that the whole world's "normal" became new.

For consumers, this Black Swan represented a completely unknown challenge, and with the intense movement of its wings, an aggressive and accelerated transformation of their behavior. As lockdowns and quarantines limited people's routines, decisions, and relationships, everything changed; the way consumers think about their priorities, search for what they want or need, how, and where they buy, and all that it implies that they get in contact with what they just acquired, either products or services, many businesses closed for good, as others

Marketing by Contingency in the Time of COVID-19: Overcoming Business Crises and Meeting Marketing Challenges. Alicia De La Peña, PhD & Juan Bernardo Amezcua Nunez, PhD (Eds.)

reinvented themselves, and others were bred into one of history's most unpredictable moments in time.

8.1 INTRODUCTION

COVID-19 has become a common and familiar term. It establishes a starting point for what has changed and could now be known as the "new normal," representing a new way of how people communicate, act, relate, and, above all, they consume. Throughout history, many events have redefined humanity's behavior: wars, natural phenomena, pandemics, and even theories of the end of the world like Y2K (year 2000), representing the ignorance of what would happen when the clock changes to January 1st, 2020. Many of these events passed through people's lives as news, rumors, or simply something distant from their reality; However, 2020 brought an event that would change every aspect of everyday life and represented people worldwide, with no exceptions. Hence the need to consider what this specific event, COVID-19, means, and represents, defined by experts as an infectious disease caused by a more recently discovered coronavirus; however, it is true in the world, there are many infectious diseases, few have had the impact that COVID-19 has had, the questions about the presence of diseases make it be considered how unusual has this event been? Who is affected? In consumer terms, was the world prepared for this unforgettable event? Is its effect extraordinary? Was it unpredictable? Based on the above questions and the impact that COVID-19 represents, it could be considered a "Black Swan."

The term "Black Swan" was more frequently used by Nassim Nicholas Taleb, mathematician-financier, researcher, and essayist, who has written his ideas on crisis scenes. As described in his most outstanding work, "The Black Swan: The Impact Of The Highly Improbable" (Penguin, 2008), the events related to the Black Swan phenomenon have a relevant impact within an environment and are difficult to predict.

Although it is true that there are many diseases and that they have occurred throughout the history of humanity, on infrequent occasions a condition has the impact that COVID-19 has had, it can be classified as a Black Swan since, in a way abruptly, wholly transformed the reality of humanity – from its implications in the field of health to how markets underwent severe transformations, which for the study of consumers, must be analyzed and studied with a comprehensive approach, particularly the change in how people react and behave in a pandemic situation, where it is likely that many people considered a possible event, but what was not quantified was the impact that COVID 19 has on the change in consumer behavior.

A Black Swan, in a consumer's mind, creates an association with past or previous experiences, trying to justify their new decision-making and buying behavior through the "triplet of opacity: the illusion of understanding, the retrospective distortion and the overvaluation of factual information and the handicap of authoritative and learned people (Figure 8.1) (Aldous, 2011)."

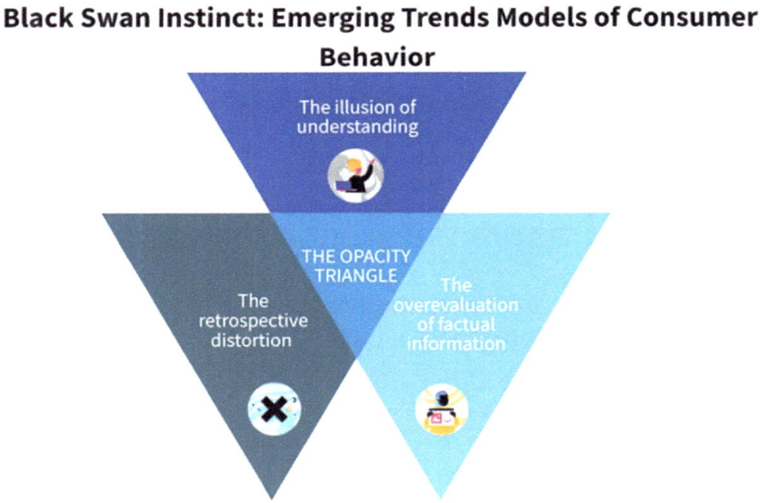

FIGURE 8.1 Emerging trends models of consumer behavior.

Source: Nassim Nicholas Taleb – The black swan: The impact of the highly improbable; Random House (2007), n.d.

Within the consumer's mind is the perception that they understand what is happening in their environment, even though their scope is limited. They know only a fraction of how complex COVID-19 is and what it entails as it is a whirlwind of processes and systems within a single word and holds a revolution for every aspect of a human being's life. It is about trying to justify human behavior and their change in conduct through retrospective distortion as if the future were an extension of the past, and foremost, taking into consideration that as consumers, the power and impact of circumstances and events that are being faced are overestimated; consumers have become COVID-19 gurus and know-it-alls regarding pandemic trends, and have even become advocates and recommenders of products and techniques on how to survive the new normal and make the best of it.

As people's autonomy has become limited, emerging trends have triggered new patterns in consumer behavior – changes in the way they shop, their lifestyle, how they live within their homes, their health and well-being, when, and where they purchase products and services, how they socialize and hold relationships – both personally and professionally, and the way they acquire and analyze information that will help them make any decisions relevant to the different roles and aspects they phase as human beings. COVID-19 accelerated changes that had been foreseen would happen in around 10 years to a couple of weeks in 2020 (Kohli et al., 2020).

8.2 EMERGING TRENDS MODELS OF CONSUMER BEHAVIOR

The year 2020 will forever be remembered as a year in which catastrophic changes happened, where emerging trends in consumer behavior models were disrupted; as many would consider, COVID-19 is a Black Swan phenomenon. It represents a disruptive event that reshaped consumer's priorities, decision-making regarding consumption and purchasing, and under uncertain scenarios and policies – transformed and modeled consumer behavior. Within the same year, three key moments emerged: (1) the World Health Organization (WHO) declared COVID-19 as a worldwide pandemic; (2) consumers prioritized the conservation and preservation of their lives by locking themselves down and avoiding physical contact, changing the way they accessed products and services necessary for their survival and well-being; and (3) consumers become anxious and willing to start living what is now considered as "the new normal" (Figure 8.2).

These trends become present as they are distanced mainly by policies and regulations at a government level, where countries began to issue guidelines that prioritized citizen's well-being through lockdowns. Although the

WHO marked the start of the pandemic on March 11th, 2020 (Ghebreyesus, 2020), countries started to make an effort to massively reduce physical contact by issuing lockdown measures and social distancing until March 26th of the same year. These measures were recommended by healthcare experts early on and were later taken into consideration by governments with a set goal: minimize the spread and impact of the virus. Extreme restrictions were imposed, and with them, a drastic change in the way consumption happened – patterns, priorities, and emerging trends, all shifted too quickly.

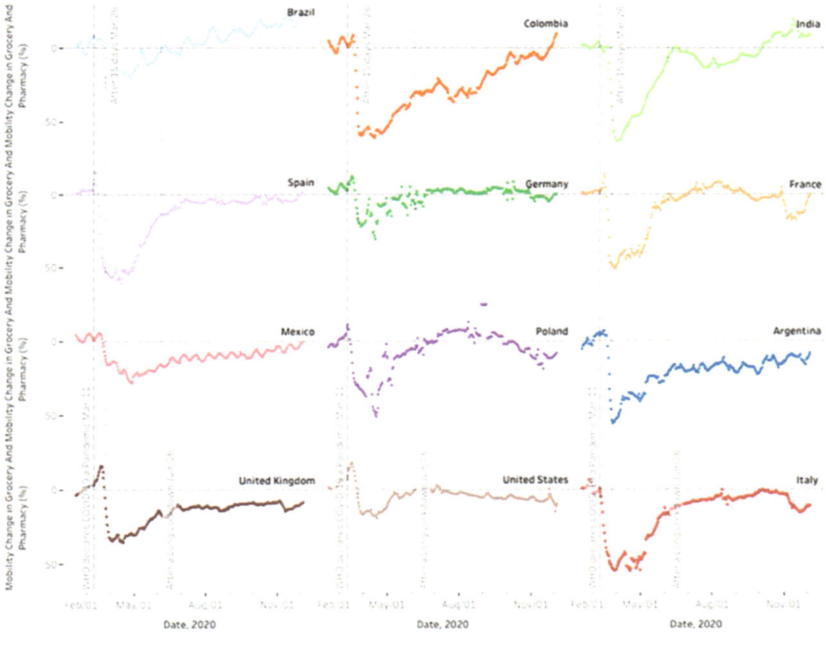

FIGURE 8.2 Change in mobility between countries with the highest number of infections (greater than 1 million infections).

Source: Grashuis et al. (2020); Johns Hopkins Coronavirus Resource Center (2020).

As consumers, and more so, as humans, the priority became safe, keeping that one love safe, and making sure their basic needs were covered. In doing so, fear kicked in through self-preservation instincts (Lindstrom, 2020). In retail, for example, an unexpected phenomenon was panic shopping, which although highly anticipated when there is little or no information on how a specific event like a war or natural disaster will impact a particular geographic area, consumers worldwide were panic-shopping toilet paper in significant quantities, valuing it almost as diamonds (Kirk and Rifkin, 2020).

Another critical factor contributing to the fear people are clinging to worldwide is the availability of information and the ease of access to it (Loxton et al., 2020). Traditional communication channels have evolved and grown into digital ones where user-generated content is widely and openly available. There is no filter to validate that the information provided is truthful. The bigger problem was the information details reaching a large population group, likely to result in emotional distress, worsening health conditions for some. As habits have changed, so have mindsets, and individuals who might have started their lockdown thinking they would have more time for themselves, are pretty much burned out and looking for a way to rewire their emotional and behavioral patterns, as the stress and dynamic pressure COVID-19 has put people under will have severe and long-lasting effects (Lindstrom, 2020).

As COVID-19 continued to spread worldwide, more investigations and analyzes started to occur, especially regarding the population's behavior and its effects. Studies and indicators determined which countries were the best and worst to live in during a pandemic event (Chang et al., 2020). The importance and relevance of these studies lie in that based on different characteristics in the classification, details, and insights can be outlined based on how consumers are acting and being affected in each of these places: what happens if consumers are living in a country where there are many or few deaths? How is mobility seen, and what is its effect/impact on day-to-day life? If there were a lot of mobility, where are people heading and why? Another essential point to consider is resilience (Chang and Jinshan, 2020) and governmental policies that may or may not dictate or model how the consumer will act (Figure 8.3).

The resilience indicator, taking into account Lockdown Severity, shows how country conditions vary based on the firmness of governmental policy. As there are more restrictions on mobility and socialization, countries with low disruption have better resilience scores, in general, than those with high disruption. Countries pictured are considered as of December 2020, where they have over 1 million infected citizens (Johns Hopkins Coronavirus Resource Center, 2020).

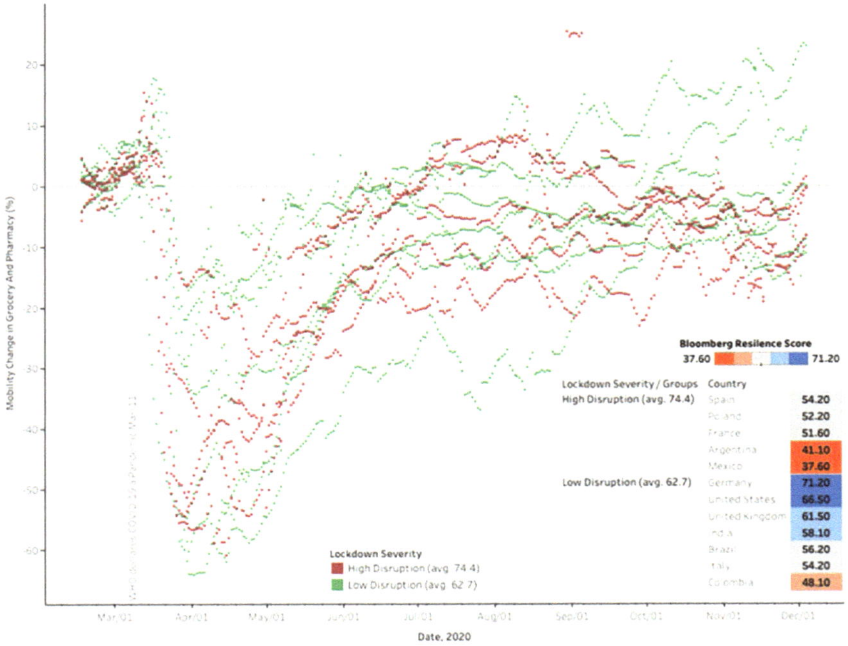

FIGURE 8.3 Change in mobility between countries with the highest number of infections (greater than 1 million infections) by Lockdown Severity Grade, according to Bloomberg.

Source: Chang et al. (2020); Grashuis et al. (2020); Johns Hopkins Coronavirus Resource Center (2020).

COVID-19 changed the way the world was working. Since the virus started in Wuhan, China, and the spread began to grow, citizens started changing their habits and the way they were living. As a pair of weeks went by, and the risk of cross-country spread increased, more awareness began to generate around the virus and the implications it would have on people's lives. Apart from the emotional exhaustion the pandemic brought, behavior in consumption changed drastically. The limitation in some dimensions brought light into others, forcing people and their providers of products and services to reinvent themselves to adapt to the "new normality." The transactional behavior of a consumer affected much their expectations, their decision-making journey, and above all, their perception of value.

8.2.1 DECISION-MAKING, BASED ON FEAR

Lockdown was one of the first actions performed to contain the spread of the virus as much as possible. Little did citizens know that as they stopped their lives from one moment to another, it would change every aspect of their lives and how they knew them. They were locked down physically and socially, geographically, and, many times, emotionally.

Because mobility was limited, the pandemic took a toll on what people could and could not do, not only regarding where they could go but what they could access, seeing that as their regular options for consumption were restricted, they were left with the opportunity to either see if their "top-of-mind" company had adapted their availability to meet their customers or look for an option which could better serve their needs. Mobility shifted completely as now, all over the world, people were expected to be at home around the clock, and perform their regular activities like school, work, exercise, and shopping from the comfort of their home (Figure 8.4) (Sheth, 2020).

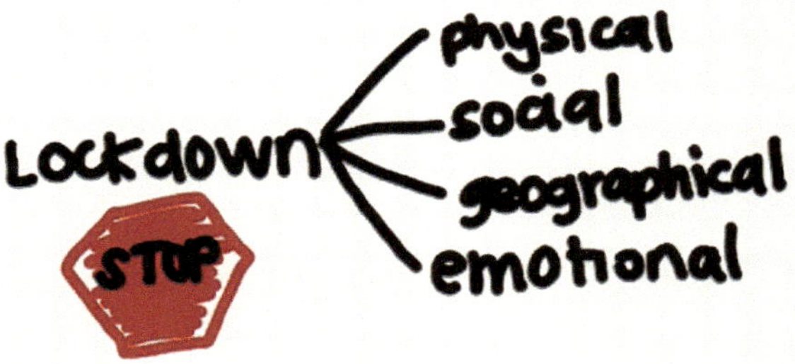

FIGURE 8.4 Mobility phases impacted by lockdown.

Source: Sheth (2020).

8.2.2 EMERGENT TRENDS IN MODELS FOR CONSUMER BEHAVIOR

As with toilet paper, hoarding has been one of the most vital consumer trends present during the appearance of COVID-19. Uncertainty world-wide kicks in people's feelings of fear and anxiety. By stocking up on specific products that mitigate Maslow's pyramid bottom level regarding basic physiological needs, the consumer's perception of security rises and suppresses fear momentarily as they purchase or get their hands on them, allowing them to feel confident that if supply were to be limited forward in time, they would be ready (Sheth, 2020). Products other than toilet paper commonly hoarded during Black Swan-type events are cleaning products – especially during a pandemic – hygiene-related items and nonperishable food. During the first few weeks of COVID-19, the main hoarded products were soap, chlorine-based cleaning products, alcohol, and specific medicines that doctors tested on their first patients (Figure 8.5).

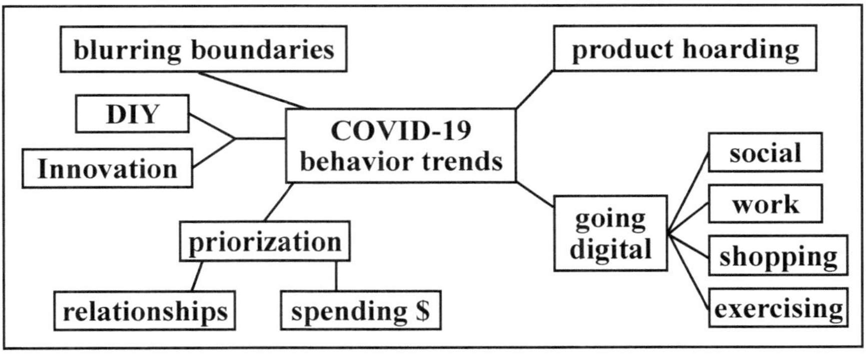

FIGURE 8.5 COVID-19 behavior trends.

Source: Sheth (2020); Kirk and Rifkin (2020).

The second greatest consumer trend was: going digital – absolutely everything commonly done or attended offline had an outstandingly quick need to migrate to online services or platforms to reach their desired (and new potential) audiences. The recommendation for physical distancing was the main trigger for this particular trend and forced individuals to fulfill their needs and routines – socializing, working, exercising, shopping – through a computer, smartphone, tablet, or television; this also brought the need for creativity to emerge at its fullest, and it did through virtual marathons and races, religious services, as well as social and live events like weddings, concerts, and even traveling (Kirk and Rifkin, 2020).

When a brand becomes a verb, as Zoom has during this particular pandemic event, it is clear the adoption of its use has become part of people's daily lives. Social media also hit new user and usage records.

Many transferred their communication and socialization solely to these platforms, creating the quickest and most impactful viral content, new communities, and essential growth for influencer profiles.

Within the socialization scope, another very impactful trend that became present during COVID-19 confinement is that humans learned to prioritize their relationships and took advantage of the development of communication technology to reunite with family and friends. More than a limitation, physical barriers became an enabler, as the opportunity and willingness to connect became more present than ever. High school and college reunions and families who had not simultaneously connected all came together through technology and the new use of apps and software that, although they had pre-COVID-19 existence, was not taken advantage of.

Throughout different countries, younger generations became shoppers for older ones. Neighbors got to know each other and more than coexisted for the first time. Regarding lifestyle, a large group of consumers settled into lockdown mode with high motivation for D.I.Y. (do it yourself) projects and an innovative spirit to face whatever challenges would come their way. This trend strongly developed as some lost their jobs and looked for new ways to get through. As hoarded products became scarce, creativity sparked towards entrepreneurship and attending to personal needs. Consumers who lived this stage became resilient quickly, as they were be foreseen scarcity and the challenge to push through even if failure presented itself. Innovation made people reinvent their companies to adapt to the new situation and find talents they didn't even know.

As COVID-19 became present in more and more countries and strong recommendations of lockdown became "the new normal," social media and platforms took over as the main channels of communication, the line between work and life practically became paper-thin. As home-office, home-school, and life adjustments based on the possibilities of each individual, a door was opened into people's homes, and with it, their life-work boundaries began to blur. Fitting in work, lifestyle activities, and habits into bedrooms, apartments, and houses made this paper-thin line even thinner, as household rooms became spaces for work rather than rest or play. This trend brought significant challenges, mainly in mental health, for many Stresses, never shutting off, having more jobs than pre-pandemic, too many roles or activities on their platter are just some of the issues people started carrying with them amidst the pandemic lockdown.

8.2.3 THE GAME RULES CHANGE FOR RETAIL

Black Swans trigger drastic changes in both consumer behavior and industries. COVID-19, compared to other pandemics and historical unexpected and life-changing events, has given people the opportunity to adapt and reinvent themselves faster than ever before because of the available technology. Because of the specific characteristics of this virus, retail was impacted all around: physical distancing empowered omnichannel buying and delivery experiences for both users and workers, product acquisition and spending prioritization evolved from "wants" to "needs," as well as looking to fulfill feelings of security and certainty over pleasure immediately.

As lockdowns started to become a new standard in countries worldwide, essential products like antibacterial soaps or wipes and face masks were expected to become high in demand by consumers, which resulted in many retailers wanting to include them in their stock and have them available for purchase, even though they weren't at all aligned with their product and product category architecture or strategy. Retailers hoped that their regular consumers, and even new ones, would become attracted to buying in their stores by having these available.

Consumer spending also shifted as options became limited to what was available online or through "essential" services such as supermarkets, convenience stores, and pharmacies. Services and experiences consumers pursued before COVID-19 lockdowns, like eating out at a restaurant, going to a movie theater or show, were not part of the "essential" services available. So industries that focused more on entertainment and non-essential activities either evolved or died.

During the COVID-19 pandemic, e-commerce became the star and savior for consumers and the retail industry. People were physically limited from shopping in brick and mortar and superstores as they did even days before the virus spread. Online shopping became a vital substitute for shopping physically as going out was avoided, and physical distancing measures impacted longer lines and limited quorums in physical spaces. This pandemic forced consumers to go online through their phones, tablets, or computers to search for the availability of products they may need to buy to assure their well-being at home. Retailers lockdowns new ways to organize, display, pack, and ship or deliver their products to their customers (Bhatti et al., 2020).

Retailers phased new challenges, especially in delivery, as the "last mile" became key to their customers' experience. Because of the shift to everything online, the exceedingly high demand in orders accompanied by the limited capacity of workers and their possibility to fulfill the number of orders they were receiving impacted the whole retail supply chain, enabling, and forcing shipping and delivery processes to transform and evolve to meet demand. Apart from having products in stock, responsive, and user-friendly online stores became a must, which should support high numbers of real-time visitors guaranteeing the correct visualization of products and check-out processes with alternative payment methods available. Home-delivery, curbside pick-up, buy online, and pick-up in-store, as well as on-demand and same-day delivery, became must-haves for retailers, improving the previously mentioned "last mile" experience they offered their customers (Figure 8.6).

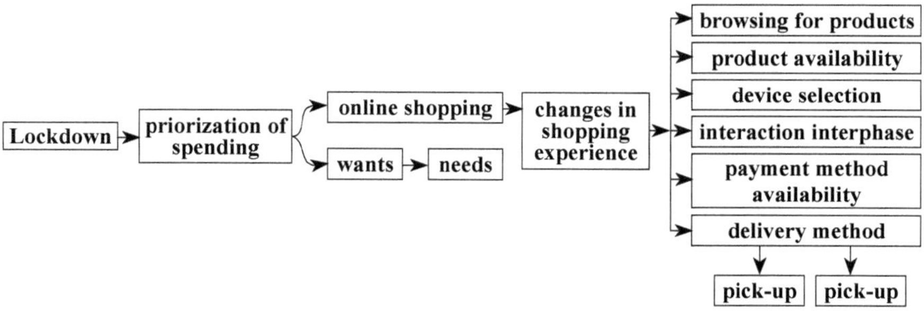

FIGURE 8.6 Experience chain towards changes in shopping experience.

Source: Bhatti et al. (2020).

8.3 CONCLUSION

What has historically been seen in a fictional book or movie has become true; the difference is that the reality that struck the world in 2020 through COVID-19 has no borders as it was not centered in a specific geographic zone as would be seen in many pandemic-related movies. The Black Swan presented itself to stay through the pressure of transforming and changing the way consumers were behaving, both as individuals and as societies. It is highly likely that from COVID-19 and on, consumers become more careful about how they decide what to buy or how they do so. A high number of e-commerce skeptics are now users of these digital platforms.

Reserve to preserve is one of the biggest trends that grew with consumers during 2020, as movies have historically portrayed pandemic events as having one goal-finding a cure and saving humanity – COVID-19 made surviving and being safe the priority. People learned, in general, that it was the starting point for a new way of working and living. Healthcare and healthy measures have become the center of every business – survival in a retail world is about more than just having good performance and income – it is about your user and customer's experience, how you operate, what channels are better and more convenient for them (to search, pay, to deliver, to get to know you), how well informed they are and how they make decisions based on the information they have. It is vital to remember; COVID-19 changed people's priorities, perception of value, and decision-making patterns – they are not the same people they were before the pandemic, as they have changed as customers, businesses need to change to meet their customer's needs. The baseline for company survival, more than people, was about who was better prepared to face every challenge, and their willingness to reinvent themselves, for which technology became a key ally, in a consumer's eyes, for the search of their priorities, the evaluation of the services and products that best fit their needs and taking in consideration the use of new payment methods. It is highly likely that as COVID-19 changed people's lives worldwide in their habits and life-changing behavior, the focus will shift towards settling in this "new normal" and trying to make sense of how to transform fear into feeling safe well-redefining value perceived by customers.

Although consumer behavior, if studied well, could be highly predictable, Black Swans are at the heart of the most unlikely events that can foresee, and their impact might be unprecedented and life changing. These types of events bring VUCA into people's lives, and with them, the change in priorities, perception of value, and a shift in behavior push providers and consumers alike to reinvent themselves and adapt to new scenarios.

KEYWORDS

- **black swan**
- **consumer behavior**
- **COVID-19**
- **trends**

REFERENCES

Aldous, D., (2011). In; Taleb, N. N., (ed.), *The Black Swan: The Impact of the Highly Improbable* (pp. 413–427). Random. In A.M.S. Notices.

Bhatti, A., Akram, H., Basit, H. M., Khan, A. U., Raza, S. M., & Naqvi, M. B. (2020). E-commerce trends during COVID-19 Pandemic. *International Journal of Future Generation Communication and Networking, 13*(2), 1449–1452.

Chang, B. R., Hong, J., & Varley. K., (2020). *The Best and Worst Places to Be in the Coronavirus Era*, 1–24. https://www.bloomberg.com/graphics/covid-resilience-ranking/?srnd=graphics-v2 (accessed on 18 February 2022).

Chang, R., & Jinshan. H., (2020). *Inside Bloomberg's Covid Resilience Ranking – Bloomberg* (pp. 1–16). Bloomberg. https://www.bloomberg.com/graphics/covid-resilience-ranking/?srnd=graphics-v2 (accessed on 18 February 2022).

Grashuis, J., Skevas, T., & Segovia. M. S., (2020). Grocery shopping preferences during the COVID-19 pandemic. *Sustainability (Switzerland), 12*(13). https://doi.org/10.3390/su12135369.

Johns Hopkins Coronavirus Resource Center, (2020). *COVID-19 Map - Johns Hopkins Coronavirus Resource Center*. Johns Hopkins Coronavirus Resource Center. https://coronavirus.jhu.edu/map.html (accessed on 18 February 2022).

Kirk, C. P., & Rifkin. L. S., (2020). I'll trade you diamonds for toilet paper: Consumer reacting, coping and adapting behaviors in the COVID-19 pandemic. *Journal of Business Research, 117*, 124–131. https://doi.org/10.1016/j.jbusres.2020.05.028.

Kohli, S., Fabius, V., & Veranen, S. M. (2020). *How COVID-19 is Changing Consumer Behavior – Now and Forever*. https://www.mckinsey.com/~/media/mckinsey/industries/retail/our insights/how covid 19 is changing consumer behavior now and forever/how-covid-19-is-changing-consumer-behaviornow-and-forever.pdf?shouldIndex=false (accessed on 11 March 2022).

Lindstrom, M., (2020). *Buyology for a Coronavirus World* (Issue 1). In Lindstrom Company.

Loxton, M., Truskett, R., Scarf, B., Sindone, L., Baldry, G., & Zhao. Y., (2020). Consumer behavior during crises: Preliminary research on how coronavirus has manifested consumer panic buying, herd mentality, changing discretionary spending and the role of the media in influencing behavior. *Journal of Risk and Financial Management, 13*(8), 166. https://doi.org/10.3390/jrfm13080166.

Sheth, J., (2020). Since January 2020 Elsevier has created a *COVID-19 Resource Center with Free Information in English and Mandarin on the Novel Coronavirus COVID-19*. The COVID-19 resource center is hosted on Elsevier Connect, the company's public news and information.

Taleb, N. N. (2007). *The Black Swan: The Impact of the Highly Improbable* (Vol. 2). Random House.

CHAPTER 9

The Role of Internal Marketing During the COVID-19 Crisis

EVA M. GUERRA-LEAL,[1] NANCY B. ORTIZ,[2] and RAÚL RUIZ[2]

[1]*EGADE Business School, Monterrey, Mexico,*
E-mail: guerra_eva@tec.mx

[2]*Marketing and Analytics Department, Tecnologico de Monterrey, Monterrey, Mexico*

ABSTRACT

The center of the marketing concept is the customer. However, within the internal marketing (IM) literature, the frontline employees' role is crucial to deliver the brand's promised service, and they are considered the first customer. Throughout well-known companies' examples and a quantitative study of employees' perceptions, we will present relevant key findings of how an IM approach could help to strengthen employees' attitudes and behaviors and recommend the employer brand to friends and family. Our findings indicate that the key elements to recommend the employer's brand (employee net promoter score (eNPS)) are the perceived organizational support (POS), the affective commitment and the organizational citizenship behavior (OCB). Additionally, in order to promote OCBs, companies should promote the company brand knowledge and normative commitment, especially during this COVID-19 crisis that has changed the employees' routines and their work-life balance.

Marketing by Contingency in the Time of COVID-19: Overcoming Business Crises and Meeting Marketing Challenges. Alicia De La Peña, PhD & Juan Bernardo Amezcua Nunez, PhD (Eds.)

9.1 INTRODUCTION

In late 2019, China authorities announced to the world that a new, very lethal, and contagious virus was found in the city of Wuhan, a virus with high levels of contagiousness for which no vaccine or cure was known. The virus quickly spread to other parts of the world, and in march 2020, the World Health Organization (WHO) declared the COVID-19 outbreak as a pandemic (WHO, 2020). Since then, millions of people have died because of the virus, and human life has changed in both the personal and the work environment.

One of the COVID-19 pandemic consequences is the change in the employees' routines and their work-life balance. Companies have faced unprecedented challenges to ensure their operational continuity. Millions of employees worldwide have turned to home-office schemes or have been forced to adapt to rigorous hygiene measures and social distancing in their workplace. Thus, an important question arises: How has this pandemic affected all those factors related to a positive organizational culture that allows companies to perform successfully in their industries?

Under this context, it was very evident that most organizations were not prepared to face a crisis of this magnitude, and they were facing too many challenges related to their employee's relationship. During this time, it was prevalent to see organizations developing new strategies to solve their internal problems, from developing new and efficient communication tools with their employees, even to adjusting jobs and schedules to make the work more flexible from home. Companies also had to face employee's low levels of commitment and learn to care about their emotional well-being in a scenario where uncertainty was predominant.

Therefore, this chapter aims to explain the basics of internal marketing (IM), why internal and external customers have the same importance, and why it is relevant to promote the company to the community, especially during crisis times as the actual COVID-19. During contingency times, it is especially relevant to promote an organizational commitment (OC), organizational values, and brand identity, give organizational support, as well as promote service-oriented citizenship behaviors. After all, we want to emphasize that although IM involves human resources (HR) variables, this is not only an HR issue; it is relevant for marketing managers because employees are a key element to deliver a high level of customer service to final customers.

This chapter will provide this IM point of view. Rafiq and Ahmed (2000) defined IM as "a planned effort using a marketing-like approach, to overcome organizational resistance to change and to align, motivate, and inter-functionally coordinate and integrate employed towards the effective implementation of corporate strategies to deliver customer satisfaction through a process of creating motivated and customer-oriented employees." In short, IM deals with applying marketing-like strategies to the company employees to motivate them to increase their results and overall performance, thus rendering better results to the firm.

Several examples of companies have implemented IM strategies, clearly understanding the importance and power of keeping their employees motivated and comfortable in their jobs. Consequently, these companies have had better results and performance due to increased employees' morale and motivation. IM is an important positive influence on corporate culture and a determinant of a company's performance because it improves communication processes with employees and allows them to internalize better corporate values (Hogg and Carter, 2000).

Google has implemented strategies to maintain their internal customer experience and, in return, they have productive, creative, and happy employees. For example, Google allows employees to use 20% of their working time on projects related to their passion and creativity. Video game rooms, silent rooms, a spa, and breakfast, lunch, and dinner are available for employees. Daycare is available for mothers with small children. The company is pet-friendly. Employees have three months' vacations. As stated in Forbes online magazine, "the company understands that employees need to be taken care of in all aspects of their lives to perform at the highest levels" (Forbes Technology Council, 2018).

Virgin is another top firm that highly values IM and has been recognized for its corporate culture. Richard Branson, the company's CEO, has established a culture oriented to make employees happy. He argues that leaders should be the ones that most favor the "employee happiness" philosophy of the company. Hence, he established hiring procedures that ensure that all the new company's leaders genuinely love and find the best in people. Branson himself said that "by taking care of employees at Virgin, everything else will take care of itself" (Krapivin, 2018). Moreover, Virgin developed the internal campaign #getonboard, with the objective that employees adopt the "Workplace" platform. The results of this campaign were successful, with an 81% adoption rate eight weeks after the implementation and 65% of weekly active users (Workplace, n.d.).

These are two examples of the importance of IM before COVID-19. However, can IM strategies help companies to adapt and perform successfully in this pandemic time? Recent studies show it might. An ongoing joint study between the MIT Sloan School of Management and Glassdoor across the Culture 500 companies show a spike in the cultural ratings of the companies due to the employees' perceptions of improved communication and integrity in the workplace as well as an improved role of the leaders and the availability of resources to do their job (Sull, et at., 2020).

This chapter will explore, through a quantitative study, the employees' of their work environment during the COVID-19 pandemic. The structure of this document is structured as follows. The following section presents a review of the literature on IM, employee net promoter score (eNPS), OC, company brand knowledge, perceived organizational support (POS), and organizational citizenship behavior (OCB). Then, we present the method and results of the quantitative study. Finally, we present a discussion of the general results and the conclusion of the study, as well as the limitations, contributions, and suggestions for future research.

9.2 LITERATURE REVIEW

IM is defined as ensuring the employees' understanding of the brand promise and their role in delivering an on-brand customer experience and satisfaction (Mosley, 2007). This definition does not empathize with the employees' role only as a 'channel to deliver a service. It goes beyond recognizing that honest and authentic brand values are born and lived inside the company. This IM approach comes from the multidisciplinary service field that integrates ideas from diverse perspectives of marketing, operations, HRs, and psychology.

Service management research focuses on how organizational variables are linked to employee attitudes and behaviors and are then reflected in customer perceptions (e.g., satisfaction, loyalty, service quality) and in financial profit. Three major perspectives on this field have emerged:

the service profit chain, customer relationship economics, organizational psychology, or human resource management (HRM) (Dean, 2004). First, the service profit chain (Heskett et al., 2008) links internal service quality (i.e., job design, human resource policies, and practices) to employees (i.e., satisfaction, productivity, and retention) and then to external customers (i.e., customer satisfaction, perceived quality, loyalty, and profits), focusing on how a positive environment for employees generates value for customers and thus leads to customer retention and profits (Dean, 2004; Towler et al., 2011). Second, the marketing view of customer relationship economics (Storbacka et al., 1994) proposes that high levels of service quality lead to high customer satisfaction and profitability levels. Finally, the organizational psychology or HRM perspective establishes a link between employees' organizational experiences concerning internal service climate (Schneider and Bowen, 1993; Schneider, 1987). This chapter adopts organizational psychology or HRM perspective as it focuses on the elements that promote employee attitudes, behaviors, and recommendations of their employer brand.

The following sections present a brief review of the literature on the antecedents of eNPS as well as a review of each of the constructs considered in the present study: company's brand knowledge (CBK), OC, POS, and OCB. The literature review aims to clarify the role of managers/employers and employees in the IM context.

9.2.1 EMPLOYEE NET PROMOTER SCORE (ENPS)

As part of this HRM perspective, this chapter uses the eNPS as the primary measure of employee engagement. The NPS measuring system was initially proposed by Reichheld (2003) to show a strong positive correlation between customer feedback and company performance. However, NPS has also been found to be a relevant metric to evaluate global employee engagement (White, 2008). According to Kaufman et al. (2013), eNPS is a metric of the degree to which an employee recommends his/her employer to qualified friends and family. It sorts respondents into three categories: promoters, passives, and detractors. Table 9.1 presents the characteristics of this scale as an engagement metric.

TABLE 9.1 eNPS Scale and Behaviors by Category

Score	Detractors 0 to 6	Passives 7 to 8	Promoters 9 to 10
Feeling	People who feel their lives have been diminished by their associations with their organization or leaders.	People who are unenthusiastic.	People who feel that their lives are enriched by their relationships with their organizations or leaders.
Satisfaction	They are dissatisfied and even dismayed by how they are treated.	People who are fairly satisfied.	Promoters go the extra mile to help customers or colleagues.
Loyalty	They and are likely to leave as soon as they find something better.	People who are fairly loyal. If a better offer comes along, they are likely to defect.	They behave like loyal employees, typically staying longer.
Word of Mouth	They frequently speak negatively about their organizations.	They rarely talk up their companies, and when they do, it's likely to be qualified.	They talk about the organization up to their friends and colleagues.

Note: Global Employee NPs is calculated as the percentage of promoters minus the percentage of detractors.

Source: Based on Kaufman et al. (2013).

9.2.2 KEY ELEMENTS TO RECOMMEND AN EMPLOYER BRAND

9.2.2.1 ORGANIZATIONAL COMMITMENT (OC)

Commitment is a connection between individual interests and a consistent activity system manifested by willingness and loyalty (Becker, 1960; Kanter, 1968). It involves a notion of membership and has predictive potential for performance, motivation, and pro-social behaviors (Brown, 1969). O'Reilly and Chatman (1986) defined it in terms of a psychological attachment, proposing an attitudinal perspective that includes three types of involvement: compliance (i.e., instrumental involvement to attain specific extrinsic rewards), identification (i.e., based on a desire for affiliation), and internalization (i.e., predicated on the congruence between individual and organizational values). In addition, they relate commitment to positive organizational consequences, such as in-role prescribed behaviors and extra-role or prosocial acts.

In line with these perspectives, Allen and Meyer (1990) proposed a multidimensional construct of OC that includes affective, continuance, and normative dimensions. According to this definition, an employee who is committed to his/her organization shows affective commitment as a positive emotion toward the organization, a continuance commitment in perceiving a high opportunity cost of leaving the organization, and a normative commitment that encompasses internalizing organizational values and having a sense of obligation toward work and organizational goals. Furthermore, Sun et al. (2013) found a direct relationship between OC (in each of the three dimensions) and word-of-mouth (WOM) intentions in the Chinese hospitality industry. In another study in China, it was found that brand community commitment (mainly by customers) had a strong effect on recommendations measured as WOM (Hur et al., 2011). However, Piercy and Carr (2020) demonstrated through a US national sample that online reviews (i.e., in websites as Indeed.com, GlassDoor.com) reveal a lot about employees' commitment and about the workplace itself. Therefore, especially for the service personnel that directly contact customers, it is imperative to create a great OC to raise their positive word of mouth about their organization (Rogers et al., 2014).

Legerstee (2013) advocates for the eNPS as a metric of employees' well-being. He found that the eNPS is a result of how employees are affective-committed to the organization and is correlated with the person-organization fit (positive) and intention to leave (negative).

Moreover, in the bank industry, Bloemer, and Odekerken-Schröder (2006) found that commitment is positively related to positive loyalty behaviors of employees, such as positive WOM. Furthermore, in that specific industry, they found that affective commitment (OC-A) has the most significant positive on positive WOM and loyalty as an intention to stay. Thus, since this study will be applied to different industries and work profiles, the first hypothesis, dividing it into one hypothesis for each commitment dimension, is defined as follows:

- **Hypothesis 1a:** OC-N has a positive effect on eNPS.
- **Hypothesis 1b:** OC-C has a positive effect on eNPS.
- **Hypothesis 1c:** OC-A has a positive effect on eNPS.

9.2.2.2 COMPANY'S BRAND KNOWLEDGE (CBK)

Company brand knowledge is defined as "the cognitive representation of distinct brand identity in employees' memory" (Terglav et al., 2016). In the context of IM, brand knowledge or a CBK is the degree of the employees' brand-relevant knowledge. CBK is the employee's cognitive awareness about the internal and external brand communication, brand values, and brand benefits and provides employees with the wherewithal to behave in the way their company's brand identity requires (Baumgarth and Schmidt, 2010). Employee's brand knowledge results in higher levels of identification and engagement. First, the latter fosters employee performance (Ngo et al., 2019) and increases employees' role clarity and affective commitment (Garas et al., 2018). Employees' OC is one of the most important determinants of favorable external representation of the workplace, according to Lages (2012), which refers to employees' verbal expressions about the workplace which shape a company's image to outsiders (Bettencourt and Brown, 2003). Therefore, it is expected that an engaged employee with knowledge about the company will recommend the brand to others. Thus, the second hypothesis is stated as follows:

- **Hypothesis 2:** CBK has a positive effect on eNPS.

9.2.2.3 PERCEIVED ORGANIZATIONAL SUPPORT (POS)

Eisenberger et al. (1986) originally described POS as a tendency to assign human-like characteristics to an organization in terms of the organization's readiness to value and reward work efforts, to meet socioemotional needs, and to care about employees' well-being. POS describes employees' belief that the organization values their contribution and cares about their well-being (Rhoades and Eisenberger, 2002). POS is the result of organizational rewards and job conditions (e.g., pay, promotions, job enrichment, and influence on organizational policies) and support programs (e.g., childcare programs, physical fitness programs, and health services) that employees believe in representing the voluntary actions of a company, as opposed to responses to external constraints (i.e., union negotiations or governmental health and safety regulations).

POS has been associated with several positive employee outcomes such as job satisfaction and life satisfaction of teachers (Bernarto et al.,

2020), affective commitment to the organization of hospital employees (Gupta, Agarwal, and Khatri, 2016), and self-determined motivation – the personal willingness or volition to do something just for the pleasure of doing it – of police officers (Gillet et al., 2012). Furthermore, POS has been found to be negatively correlated with some undesired outcomes of an employee-organization relationship. In a meta-analytic study of organizational support theory, Kurtesis et al. (2017) found that POS was negatively related to job stress, burnout, and work-family conflict. Also, Rhoades, Eisenberger, and Armeli (2001) suggested a negative relationship between POS and employee turnover in retail employees. As seen, POS has been related to positive outcomes, favorable both for the employee and for the organization, that essentially result in higher employee satisfaction.

Therefore, when employees are part of a service environment and receive organizational support, their behaviors and attitudes are oriented toward better service results, such as commitment to service, which is in turn strongly related to consumer loyalty and positive 'word of mouth' towards the company (Liang, Tseng, and Lee, 2010), thus contributing to increase the eNPS (Figure 9.1). Then, the third hypothesis is stated as follows:

- **Hypothesis 3:** POS has a positive effect on eNPS.

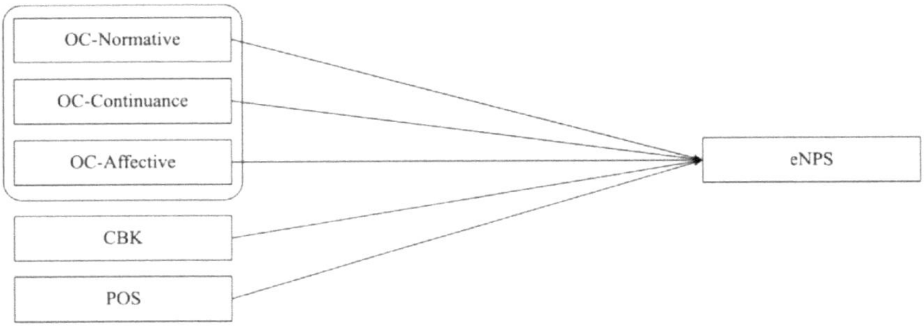

FIGURE 9.1 The model based on direct relationships.

9.2.3 MEDIATION EFFECT OF OCB OVER EMPLOYEE NPS

OCB consists of discretionary work behaviors not directly or explicitly recognized by the formal reward system that promotes the effective functioning of organizations (Organ, 1988). Podsakoff et al. (1990) demonstrated that OCB is explained by five dimensions that are contextual behaviors not explicitly prescribed by the organization: *conscientiousness* describes employees exceeding what is minimally required by the company and/or their jobs (e.g., attendance, punctuality, and rule compliance); *sportsmanship* implies employees' willingness to ignore minor inconveniences or impositions that may arise in daily work activities; *civic virtue* refers to employees' constructive participation and political involvement in an organization and how much they care about the company; *courtesy* entails employee actions directed toward the prevention of problems encountered by work associates, and *altruism* refers to employee behaviors directed toward helping or cooperating with other employees in face-to-face interactions on organizationally relevant issues. In the organizational behavior literature, OCB has been linked to organizational effectiveness (Organ, 1988) and has been found to affect customer loyalty and performance measures in service industries (Barroso-Castro et al., 2004). The main consequences of OCB are organizational profitability, performance, and success and are associated with managers' judgments about evaluations of performance, pay raises, reward allocating, and promotions. In addition, OCB is negatively related to employee turnover, turnover intentions, and absenteeism but positively related to customer satisfaction (Podsakoff et al., 2000).

Moreover, even though OC represents only a psychological attachment, attitude, or willingness to exert a substantial effort, and not an actual behavior, Chahal and Mehta (2010); Meyer et al. (2002); and O'Reilly and Chatman (1986) argue that OCB – which is, indeed, an actual employee behavior-, is a relevant outcome of commitment.

This relationship between OC and OCB has been widely demonstrated (Cetin et al., 2015; Al-Jabari and Ghazzawi, 2019) from local governments workers in Africa (Obedgiu et al., 2017), employees in a National Hydroelectric Corporation in India (Bakhshi et al., 2011), hotel workers in Iraq (Massoudi et al., 2020), to academics in Indonesia (Djaelani et al., 2020; Paramaartha et al., 2019).

However, there is no consensus about which OC dimensions have a greater impact on OCB. In a study in a telecommunication organization, Mohamed and Anisa (2012) found that normative commitment and continuance commitment have a significant impact on OCB, but affective commitment had not a significant impact on OCB.

According to Bag (2020), employees' commitment is a key piece for the success of any organization, and one psychometric tool to measure the effects of commitment is the Net Promoter Score because employees can also act as promoters, passives, and detractors. "The greater an employee's engagement, the more likely he or she is to —go the extra mile and deliver excellent on-the-job performance" (Bag, 2020, p. 66). This implies that a committed employee will recommend his/her company because he/ she is willing to do his/her best for the company. Moreover, according to Al-Jabari and Ghazzawi (2019), one of the major trends in the study of OC in an organizational context is to explore the role that commitment plays during organizational challenges, such as the COVID-19 actual situation. Therefore, the fourth hypotheses are established as follows, according to each of the OC dimensions (i.e., normative, continuance, and affective commitment):

- **Hypothesis 4a:** The relationship between OC-N and eNPS is mediated by OCB.
- **Hypothesis 4b:** The relationship between OC-C and eNPS is mediated by OCB.
- **Hypothesis 4c:** The relationship between OC-A and eNPS is mediated by OCB.

Additionally, Podsakoff et al. (2000) suggest four types of antecedents of OCB, all of which contribute to the development of an OCB environment: individual or employee characteristics (e.g., satisfaction, and commitment), task characteristics (e.g., challenging non-repetitive tasks), organizational characteristics (e.g., formalization, flexibility, supervisor or staff support), and leadership behaviors (e.g., listen, and give feedback, motivate, empower, reward). In this sense, the employee's knowledge about the organizational characteristics such as company communication, organizational goals, company values, and customer value proposition leads to positive OCBs (Van Nguyen et al., 2019). Therefore, evidence suggests OCB is a consequence of CBK. In fact, brand knowledge is a predictor of OCB (Bataineh et al., 2017). Satisfied employees are expected to spread positive comments about their organization to others (Bravo et al., 2017). Thus, according to Lages (2012), once an employee knows the company's characteristics, works for it, and behaves in a proactive manner, then he/she would recommend the company. Hence, the fourth hypothesis is as follows:

- **Hypothesis 5:** The relationship between CBK and eNPS is mediated by OCB.

Rhoades and Eisenberger (2002) suggest that employees who work in an organization that voluntarily supports them – high POS – in opposition to companies being forced by labor laws and/or union negotiations, will develop a positive job-related affect derived from socioemotional needs met by the company efforts, therefore rendering high levels of job satisfaction. One possible explanation for this positive relationship is given by Caesens et al. (2017), who suggest that employees feel more satisfied and with the highest level of well-being in high POS conditions because they feel less dehumanized; they feel they are treated like persons rather than just as a "number." In addition, Organ and Ryan (1995) found a positive relation between OCB and job satisfaction. Their meta-analytic review of OCB didn't establish a clear direction between OCB and job satisfaction. However, the possibility exists that employees who show positive citizenship behaviors are also satisfied with their jobs. Thus, it may be a virtuous circle: employees with high job satisfaction will be inclined to display higher levels of OCB's, resulting in higher job satisfaction. Moreover, job satisfaction has been positively related to happiness: employees who are satisfied in their jobs show higher levels of global happiness, a person's satisfaction with its own life, than unsatisfied employees (Satuf et al., 2018).

In the tourism industry, customer satisfaction is of utmost importance for any organization. According to Abubakar and Mavondo (2014), Thus, a positive interaction between customer-employees and customer's happiness with the overall experience will lead to higher customer satisfaction and positive WOM. The importance of employees is evident for a high-quality service: employees are in charge of several tasks that directly impact the experience, such as service in restaurants, service at the front desk, and in several other areas of the organization. Excellent service from employees will make customers feel satisfied with the service. The authors also suggest that happiness is also related to higher customer satisfaction and positive WOM. They argue that happiness is a deep feeling of satisfaction that antecedes customer satisfaction and suggest that a destination should provide the services and personnel to make customers happy through high-level services and great experiences (Abubakar and Mavondo, 2014).

These findings may suggest that any organization, not only a touristic destination, should provide the means for both employees and customers to be happy and satisfied. If employees are happy in the organization, if they feel that the organization treats them as people, not as numbers, and cares about them (POS), and if the organization rewards them for exhibiting positive behaviors towards the organization (OCB), it is very likely that those employees will be very satisfied with the company and that, therefore, they will speak positively about it to others (WOM). Therefore, the sixth hypothesis is defined as follows:

- **Hypothesis 6:** The relationship between POS and eNPS is mediated by OCB.

Hence, the existing model that includes all these variables is represented in Figure 9.2.

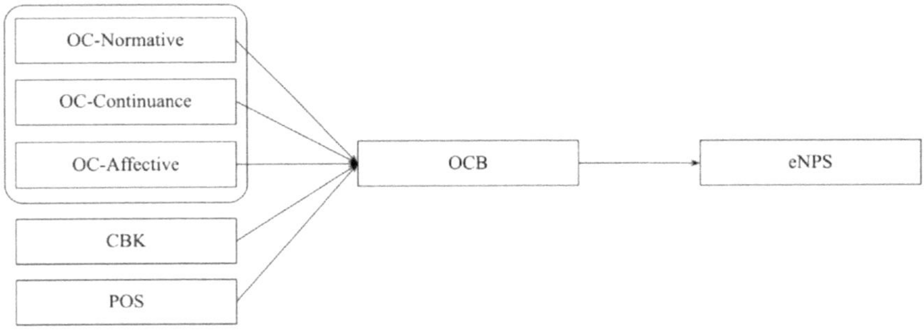

FIGURE 9.2 The model based on the mediation effect of OCB.

This study uses a quantitative methodology to test a theoretical model about the IM perceptions of employees working for different organizations. The method and results of our study are explained in further sections.

9.3 METHOD

This study is focused on a variety of industries and managerial marketing practices because the focus is on the employee's perception of their companies during the COVID-19 contingency between November 5 and November 16, 2020. Through a snowball sample, the researchers' shared the Qualtrics link to people with a current employment relationship with a company. A total of 286 online questionnaires were collected. However, the final research sample consists of 191 employees who completed all the questions. Among the final usable questionnaires, 90.5% of the respondents were from Mexico; 78.0% were full-time employees; 53.4% were female; 50.8% have postgraduate studies; 66.0% work for a large company (250+ employees); 40.3% have less than five years working for the organization; 49.7% have 1–5 years in their position; 30.4% were mid-level managers; with an average age of 41.12 years (see Table 9.2).

TABLE 9.2 Sample Composition

Gender		**Actual Laboral Status**	
Masculine	45.00%	Full-time employee	78.00%
Feminine	53.40%	Partial time employee	9.40%
Prefer not to answer	1.60%	Owner	9.90%
Residence Place		Student Trainee	2.70%
Mexico	90.50%	**Organizational Tenure**	
Latin America	7.40%	Less than 5 years	40.30%
North America	1.60%	5 to 10 years	20.40%
Other	0.50%	11 to 20 years	20.50%
Educational Level		21 to 30 years	14.10%
Postgraduate Degree	50.80%	More than 30 years	4.70%
University	42.90%	**Position**	
Highschool	6.30%	C-Level	17.30%
Age		Mid-level managers	30.40%
Less than 25 years old	9.40%	Academic	15.70%
26 to 40 years	40.80%	Administrative Staff	13.10%
41 to 55 years	37.20%	Customer service / sales	3.10%
More than 55 years	12.60%	Software developer	2.10%
Organization Size		Other	18.30%
Less than 10 employees	9.40%	**Position Tenure**	
11 to 30 employees	12.60%	Less than 1 year	14.10%
31 to 100 employees	5.70%	1 to 5 years	49.70%
101 to 250 employees	6.30%	5 to 10 years	16.80%
Less than 250 employees	66.00%	More than 10 years	19.40%

The following variables were used in this study: OC, CBK, POS, OCB, and eNPS. Table 9.3 shows the questions used in the instrument.

TABLE 9.3 Description of Variables Used in the Study

Construct	Items	Cronbach Alpha
Normative Commitment		**0.841**
OC-N1	I feel an obligation to remain with my current employer.	
OC-N2	Even if it were to my advantage, I do not feel it would be right to leave my organization now.	
OC-N3	I would feel guilty if I left this organization now.	
OC-N4	This organization deserves my loyalty.	
OC-N5	I would not leave my organization right now because I have a sense of obligation to the people in it.	
OC-N6	I owe a great deal to my organization.	
Continuance Commitment		**0.853**
OC-C1	It would be very hard for me to leave my organization right now, even if I wanted to.	
OC-C2	One of the few negative consequences of leaving this organization would be the scarcity of available alternatives.	
OC-C3	Right now, staying with my organization is a matter of necessity as much as desire.	
OC-C4	I feel that I have too few options to consider leaving this organization.	
OC-C5	If I had not already put so much of myself into this organization, I might consider working elsewhere.	
OC-C6	Too much of my life would be disrupted if I decided I wanted to leave my organization now.	

TABLE 9.3 *(Continued)*

Construct	Items	Cronbach Alpha
Affective Commitment		**0.928**
OC-A1	I would be very happy to spend the rest of my career with this organization	
OC-A2	I really feel as if this organization's problems are my own.	
OC-A3	I feel a strong sense of belonging to my organization.	
OC-A4	I feel emotionally attached to this organization.	
OC-A5	I feel like part of the family at my organization.	
OC-A6	This organization has a great deal of personal meaning for me.	
Company's Brand Knowledge		**0.860**
CBK1	I am familiar with our brand communication (e.g., magazines, Internet, Exhibitions, etc.)	
CBK2	I am aware of the goals we try to achieve through the brand.	
CBK3	I am well informed about the values represented by the brands.	
CBK4	I understand how our customers can benefit from our brand.	
CBK5	I am familiar with our brand style guide.	
CBK6	I know which attributes of our brand differentiate us from our competitors.	
CBK7	I know how to comport myself so as to present our brand to customers positively.	
Perceived Organizational Support		**0.950**
POS1	My organization takes pride in my accomplishments.	
POS2	My organization really cares about my well-being.	
POS3	My organization values my contributions to its well-being.	

TABLE 9.3 *(Continued)*

Construct	Items	Cronbach Alpha
POS4	My organization strongly considers my goals and values.	
POS5	My organization shows little concern for me. (R)	
POS6	My organization is willing to help me if I need a special favor.	
Organizational Citizenship Behavior		**0.884**
OCB1	I am assisting co-workers to deliver high-quality customer oriented service.	
OCB2	To serve my customers, I volunteer for things that are not required.	
OCB3	I make innovative suggestions to improve customer service.	
OCB4	I expend considerable energy to come up with creative ways to assist customers facing problems.	
OCB5	I attend functions that are not required, but that help customer service.	
OCB6	I exchange ideas with colleagues on how to improve customer service.	
OCB7	I deal restlessly with customer problems until they are resolved.	
Employee Net Promoter Score		**N/A**
eNPS1	How likely do you recommend to other people to seek employment at the company you work for?	

9.3.1 ORGANIZATIONAL COMMITMENT (OC)

The OCQ scale (Meyer and Allen, 1997) measures OC as a multidimensional construct (i.e., affective, continuance, and normative dimensions) using a 5-point Likert-type scale ranging from 1 (strongly disagree) to 5 (strongly agree). First, the normative dimension consists of six items (α = 0.841). Second, the continuance dimension consists of six items (α = 0.853). Third, the affective dimension consists of six items (α = 0.928).

9.3.2 COMPANY'S KNOWLEDGE (CBK)

The CBK instrument was adapted from the internal brand knowledge scale of Baumgarth et al. (2010). The instrument includes seven items (α = 0.860), on a 5-point Likert-type scale ranging from 1 (strongly disagree) to 5 (strongly agree).

9.3.3 PERCEIVED ORGANIZATIONAL SUPPORT (POS)

The POS scale consists of six items (α = 0.950). Eisenberger et al. (2001) adapted the scale from Eisenberger et al. (1986) on a 5-point Likert-type scale ranging from 1 (strongly disagree) to 5 (strongly agree).

9.3.4 ORGANIZATIONAL CITIZENSHIP BEHAVIOR (OCB)

OCB was measured using seven items (α = 0.884) from Dimitriades (2007), who presented a customer orientation or service orientation form of OCB adapted from Morrison's (1996) work regarding traditional OCB dimensions. Although it is not the most widely used OCB scale, it is adapted to service-oriented behaviors in service contexts. The participants responded on a 5-point Likert-type scale ranging from 1 (strongly disagree) to 5 (strongly agree).

9.3.5 EMPLOYEE NET PROMOTER SCORE (ENPS)

eNPS consists of a single item about how an employee recommends a company for a friend or colleague to work in it. The item is "How likely do you recommend to other people to seek employment at the company you work for?" It consists of a 0–10 scale and allows to classify respondents as promoters (score 9–10), passives (score 7–8), and detractors (score 0–6). It is a measure of employment WOM.

9.3.6 CONTROL VARIABLES

In addition to the previously specified variables, this study considers some control variables: gender, residence place, education level, age, job position, working status, tenure (in the organization, position), and organization size.

9.4 RESULTS

This section presents descriptive data on the variables composing the proposed model and the statistical analyzes. For the eNPS variable, the results were 66% promoters, 20.9% passives, and 13.1% detractors. There were no statistically significant differences between gender, residence place, education level, age, job position, working status, tenure (in the organization or the position), and organization size. From this point, we take into account the individual metric score.

9.4.1 DESCRIPTIVE ANALYSIS

Table 9.4 displays the means, standard deviations, reliabilities, and correlation coefficients for the study variables and indicates the significant relationships between these variables.

The highest value for each variable is five, and the lowest value is one, except for eNPS (1–10). With regard to the antecedents to eNPS, results show that mean values above the overall means on OCB, CBK, and POS, whereas OC-C exhibits the lowest mean. These results indicate the overall employees' perceptions during the COVID-19 crisis.

9.4.2 HYPOTHESIS TESTING

Using linear regression modeling in SPSS, results indicated that two of the direct hypothesized relationships were supported (Table 9.5). Furthermore, the parameter estimates indicated that POS was positively related to eNPS (β =.561; p <.000), supporting Hypothesis 3. Consistent with Hypothesis 1c, OC-A was positively related to eNPS (β =.284; p <.000). The R square indicates that 0.622 of the overall variances of eNPS is explained by these two variables: POS and OC-A.

However, hypotheses 1a and 1b were not supported. In Hypothesis 1a, the predicted positive direct relationship between OC-N and eNPS was not found, although they have a significant and positive correlation (0.492; p <.01).

TABLE 9.5 Results of the Path Analysis of the Model

	Hypothesis		**Unstandardized Coefficient**	**Std. Error**	**Standardized Coefficient**	**P**	**R^2**	**R^2Adj**
H1a	OC-N	→eNPS	–0.070	–1.116	0.266	0.504	0.626	0.622
H1b	OC-C	→eNPS	–0.069	–1.405	0.162	0.819		
H1c	OC-A	→eNPS	0.530	0.121	0.284	0.000***		
H2	CBK	→eNPS	0.038	0.654	0.514	0.585		
H3	POS	→eNPS	1.154	0.133	0.561	0.000***		
Mediation	OCB	→eNPS	0.523	0.222	0.169	0.020**	0.028	0.023
H4a	OC-N	→OCB	0.152	0.045	0.231	0.001***	0.316	0.304
H4b	OC-C	→OCB	–0.058	–0.769	0.443	0.662		
H4c	OC-A	→OCB	0.069	0.754	0.452	0.438		
H5	CBK	→OCB	0.375	0.070	0.369	0.000***		
H6	POS	→OCB	–0.007	–0.078	0.938	0.496		
Control	**Education Level**	**→OCB**	**0.267**	**0.068**	**0.239**	**0.000*****		

**$p < 0.001$.
**$p < 0.01$.
*$p < 0.05$.

Hypothesis 1b was not statistically supported either; OC-C is not an antecedent of eNPS. Finally, in Hypothesis 2, although CBK and eNPS have a significant positive correlation (0.521; $p < .01$), CBK is not a direct antecedent to eNPS. The control variables that were included in the model were gender, residence place, education level, age, job position, working status, tenure (in the organization, in the position), and organization size. None of these control variables were significant in the first model.

9.4.3 MEDIATION ANALYSIS

To test for mediation, the B-K mediation process (Ato and Vallejo, 2011; Judd and Kenny, 1981; Preacher and Hayes, 2004) was performed. This procedure evaluates whether the indirect effect of the independent variable on the dependent variable through the mediator variable is significant. We calculated the effect of OCB on NPS, resulting in statistically significant (0.521; $p < .05$), and then the effects of CBK, OC, and POS over OCB. The parameter estimates indicated that OC-N was positively related to OCB ($\beta = .231$; $p < .001$), supporting Hypothesis 4a. Consistently, Hypothesis 5, CBK was positively related to OCB ($\beta = .369$; $p < .000$). However, hypotheses 4b, 4c, and 6 were not supported. First, in Hypothesis 4b, although OC-C and OCB have a significant and positive correlation (0.156; $p < .05$), OC-C is not a direct antecedent to OCB or eNPS. Second, in Hypothesis 4c, there is no relationship between OC-A and Employer OCB, even when they have a significant and positive correlation (0.388; $p < .01$). Finally, Hypothesis 6 was not statistically supported either; POS is not an antecedent of OCB. The control variables included in the model were gender, residence place, education level, age, job position, working status, tenure (in the organization, in the position), and organization size. However, the only significant variable was education level. The R square indicates that 0.304 of the overall variances of OCB is explained by CBK, OC-N, and education level.

These findings indicate that OCB is a full mediator (it has a suppression effect) of the CBK-NPS and between OC-N and NPS relationships, as is indicated in Figure 9.3.

The discussion of the main findings of this quantitative study are presented in further section.

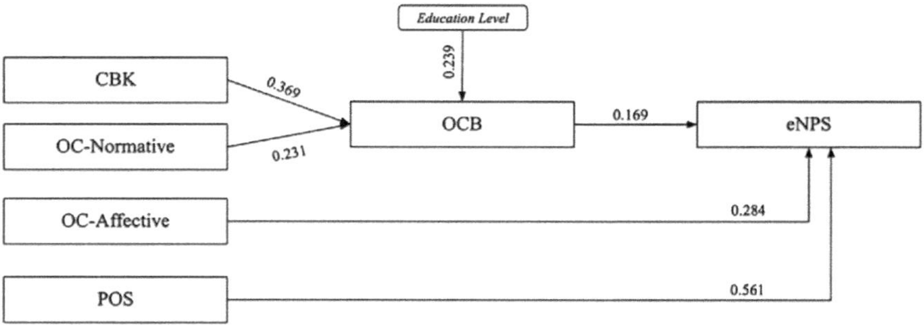

FIGURE 9.3 The final model.

9.5 DISCUSSION

This study had the objective of testing the direct and the mediating effect of some key elements to recommend an employer brand, namely OC, CBK, POS, and OCB with eNPS, a relevant measure of employee engagement. The study was directed to men and women working full-time or part-time in companies mainly in Mexico and having different education levels, ages, positions, and organization tenure.

Our results suggest that eNPS or the recommendation of a company as a "great place to work" is a consequence of the affective OC or a positive emotion toward the organization where the people want to stay working for (Allen and Mayer, 1990). This is consistent with the findings of Bloemer and Odekerken-Schröder (2006); and Legerstee (2013), in which the affective dimension had the strongest relation to WOM and eNPS, respectively. Results also show that the level of organizational support perceived by employees, or POS, is also an antecedent of a positive eNPS towards the company, an idea consistent with Abubakar and Mavondo (2014) suggestion that an organization must design an experience that includes customer-oriented employees and a great overall experience – issues clearly related to POS-, to deliver exceptional service, that result in customer satisfaction and positive WOM. As such, we can suggest that these positive recommendations are also a consequence of the POS or the general belief that their work organization values their contribution and cares about their well-being (Rhoades and Eisenberger, 2002).

These two factors are particularly relevant during contingency times because of the job uncertainty that predominates as a consequence of COVID-19. Today employees, more than ever, want to feel safe and stable and want to believe that the organization they work for cares about them and supports them. This means that the more an organization cares about their employees, the more they will have an affective OC because they will perceive that the organization values them and is also aware of their personal well-being. Therefore, organizational support and affective OC can be considered two very relevant antecedents of employees' happiness and well-being, therefore resulting in positive comments.

Additionally, our study confirms that company brand knowledge and normative OC are indirectly related to the employee's recommendation throughout the OCB. This is a relevant finding because it supports the idea that it is not enough for a company to give employees information about the company or its brands, or information about the organizational values, goals, and other formal information, to elicit positive recommendations from its employees. For this to occur, the organization needs to build a positive work environment that depends on how well the company brand identity is communicated (Terglav et al., 2016) and how well the organizational values and goals are promoted (Allen and Mayer, 1990). That is, the organization needs to work on developing positive citizenship behaviors in its employees to make this formal information useful towards the goal of eliciting positive comments from employees. This is consistent with Mohamed and Anisa (2012) findings of how normative commitment had not a significant impact on OCB, while affective commitment did not. A surprising finding is that the continuance commitment was not statistically significant for the model; maybe because of the three dimensions of OC, it is more transactional.

During these contingency times, employees will talk and share about their experiences on the companies they work for, good, and bad, and with the digital communications tools that are available nowadays, their voice will be spread very fast. Any company will like those employees share good experiences from their workplace, as external marketing. Thus, developing internal strategies to promote the brand, building positive brand knowledge as a "great place to work," and a normative commitment from employees will promote employee's recommendations to attract the best talent available and keep actual talent loyal to the company.

9.6 CONCLUSION

Since COVID-19 contingency has changed workplace dynamics, marketing managers and practitioners must implement IM strategies that enable them to face the new challenges. Thus, despite economic resources being scarce during the contingency, companies should invest in IM programs to attend to each of these challenges, improve their performance, and increase their economic results. In this sense, employees are internal customers, and they are as important as external ones because of their impact on organizational performance.

This chapter contributes to IM literature by establishing important relationships on the antecedents of employee engagement and employer recommendation. Despite our study limitations, such as the limited sample size, the convenience sample, and the particular Latin American context of the sample, this study adds to the existing literature by identifying some usable direct and indirect antecedents of eNPS, a topic that has not been fully studied by researchers. The relevance of our findings for companies is clear. The variables shown to be direct or indirect antecedents of eNPS, namely affective and normative OC, POS, and customer brand knowledge, are all in control of the company. Therefore, an organization must focus on developing those conditions that contribute to improving these elements and et more satisfied and happy employees that talk positively to others about the company, especially in these pandemic times.

KEYWORDS

- **company's brand knowledge**
- **employee net promoter score**
- **internal marketing**
- **organizational citizenship behavior**
- **organizational commitment**
- **perceived organizational support**

REFERENCES

Abubakar, B., & Mavondo. F., (2014). Tourism destinations: Antecedents to customer satisfaction and positive word-of-mouth. *Journal of Hospitality Marketing & Management, 23*(8), 833–864.

Al-Jabari, B., & Ghazzawi. I., (2019). Organizational commitment: A review of the conceptual and empirical literature and a research agenda. *International Leadership Journal, 11*(1), 78–119.

Allen, N. J., & Meyer. J. P., (1990). The measurement and antecedents of affective, continuance and normative commitment to the organization. *Journal of Occupational Psychology, 63*(1), 1–18.

Ato, M., & Vallejo, G. (2011). The effects of third variables in psychological research. *Anales de Psicologia, 27*(2), 550–561.

Bag, P., (2020). Employees commitment fostering organizational outcomes. *International Journal of Trend in Research and Development, 5*(2), 63–66.

Bakhshi, A., Sharma, A. D., & Kumar. K., (2011). Organizational commitment as predictor of organizational citizenship behavior. *European Journal of Business and Management, 3*(4), 78–86.

Barroso-Castro, C., Armario, E. M., & Ruiz. D. M., (2004). The influence of employee organizational citizenship behavior on customer loyalty. *International Journal of Service Industry Management, 15*(1), 27–53.

Bataineh, A. Q., Alfalah, T. F., Falah, J. F., & Idris. M. I., (2017). The effect of employee-based brand equity on organizational citizenship behavior: The mediating role of job satisfaction. *International Journal of Academic Research in Business and Social Sciences, 7*(2), 423–436.

Baumgarth, C., & Schmidt. M., (2010). How strong is the business-to-business brand in the workforce? An empirically-tested model of 'internal brand equity' in a business-to-business setting. *Industrial Marketing Management, 39*(8), 1250–1260.

Becker, H., (1960). Notes on the concept of commitment. *American Journal of Sociology, 66*(1), 32–40.

Bernarto, I., Bachtiar, D., Sudijbo, N., Surayawan, I., Purwanto, A., & Asbari. M., (2020). Effect of transformational leadership, perceived organizational support, job satisfaction toward life satisfaction: Evidences from Indonesian teachers. *International Journal of Advanced Science and Technology, 29*(3), 5495–5503.

Bettencourt, L., & Brown. S., (2003). Role stressors and customer-oriented boundary-spanning behaviors in service organizations. *Journal of the Academy of Marketing Science, 31*, 394–408.

Bloemer, J., & Odekerken-Schröder, G., (2006). The role of employee relationship proneness in creating employee loyalty. International *Journal of Bank Marketing, 24*(4), 252–264.

Bravo, R., Buil, I., De Chernatony, L., & Martínez. E., (2017). Managing brand identity: Effects on the employees. *International Journal of Bank Marketing.*

Brown, M. E., (1969). Identification and some conditions of organizational involvement. *Administrative Science Quarterly, 14*(3), 346–355.

Caesens, G., Stinglhamber, F., Demoulin, S., & De Wilde, M., (2017). Perceived organizational support and employees' well-being: The mediating role of organizational dehumanization. *European Journal of Work and Organizational Psychology, 26*(4), 527–540.

Cetin, S., Gürbüz, S., & Sert. M., (2015). A meta-analysis of the relationship between organizational commitment and organizational citizenship behavior: Test of potential moderator variables. *Employee Responsibilities and Rights Journal, 27*(4), 281–303.

Chahal, H., & Mehta. S., (2010). Antecedents and consequences of organizational citizenship behavior (OCB): A conceptual framework in reference to health care sector. *Journal of Services Research, 10*(2), 25.

Dean, A. M., (2004). Links between organizational and customer variables in service delivery: Evidence, contradictions and challenges. *International Journal of Service Industry Management, 15*(4), 332–350.

Dimitriades, Z. S., (2007). The influence of service climate and job involvement on customer-oriented organizational citizenship behavior in Greek service organizations: A survey. *Employee Relations, 29*(5), 469–491.

Djaelani, A., Sanusi, A., & Trianmanto. B., (2020). Spiritual leadership, job satisfaction, and its effect on organizational commitment and organizational citizenship behavior. *Management Science Letters, 10*(16), 3907–3914.

Eisenberger, R., Armeli, S., Rexwinkel, B., Lynch, P. D., & Rhoades. L., (2001). Reciprocation of perceived organizational support. *Journal of Applied Psychology, 86*(1), 42.

Eisenberger, R., Huntington, R., Hutchison, S., & Sowa. D., (1986). Perceived organizational support. *Journal of Applied Psychology, 71*(3), 500–507.

Forbes Technology Council, (2018). *13 Reasons Google Deserves Its 'Best Company Culture' Award.* Forbes. https://www.forbes.com/sites/forbestechcouncil/2018/02/08/13-reasons-google-deserves-its-best-company-culture-award/?sh=284308ef3482 (accessed on 18 February 2022).

Garas, S. R. R., Mahran, A. F. A., & Mohamed. H. M. H., (2018). Internal corporate branding impact on employees' brand supporting behavior. *Journal of Product & Brand Management, 27*(1), 79–95.

Gillet, N., Huart, I., Colombat, P., & Fouquerau. E., (2012). Perceived organizational support, motivation, and engagement among police officers. *Professional Psychology: Research and Practice, 44*(1), 46–55.

Gupta, V., Agarwal, U. A., & Khatri, N. (2016). The relationships between perceived organizational support, affective commitment, psychological contract breach, organizational citizenship behaviour and work engagement. *Journal of Advanced Nursing, 72*(11), 2806–2817.

Heskett, J. L., Jones, T. O., Loveman, G. W., Sasser, W. E., & Schlesinger. L. A., (2008). Putting the service-profit chain to work. *Harvard Business Review,* 118–129.

Hogg, G., & Carter. S., (2000). Employee attitudes and responses to internal marketing. *Internal Marketing: Directions for Management* (pp. 109–124). Routledge, London.

Hur, W. M., Ahn, K. H., & Kim. M., (2011). Building brand loyalty through managing brand community commitment. *Management Decision, 49*(7), 1194–1213.

Judd, C. M., & Kenny. D. A., (1981). Process analysis: Estimating mediation in treatment evaluations. *Evaluation Review, 5*(5), 602–619.

Kanter, R. M., (1968). Commitment and social organization: A study of commitment mechanisms in utopian communities. *American Sociological Review, 33*(4), 499–517.

Kaufman, J., Markey, R., Burton, S. D., & Azzarello. D., (2013). *Who's Responsible for Employee Engagement.* Bain Insights Rep.

Krapivin, P., (2018). *Sir Richard Branson's Five Billion Reasons to Make Your Employees and Candidates Happy*. Forbes. https://www.forbes.com/sites/pavelkrapivin/2018/07/09/sir-richard-bransons-5-billion-reasons-to-make-your-employees-candidates-happy/?sh=1d3c94f66710 (accessed on 18 February 2022).

Kurtessis, J., Eisenberger, R., Ford, M., Buffardi, L., Stewart, K., & Adis. C., (2017). Perceived organizational support: A meta-analytic evaluation of organizational support theory. *Journal of Management, 43*(6), 1854–1884.

Lages, C. R., (2012). Employees' external representation of their workplace: Key antecedents. *Journal of Business Research, 65*(9), 1264–1272.

Legerstee, T., (2013). *Asking Employees' the Ultimate Question': Developing the Employee Promoter Score*. Erasmus Universiteit. https://api.semanticscholar.org/CorpusID:44006815 (accessed on 18 February 2022).

Liang, R. D., Tseng, H. C., & Lee. Y. C., (2010). Impact of service orientation on frontline employee service performance and consumer response. *International Journal of Marketing Studies, 2*(2), 67–76.

Massoudi, A. H., Jameel, A. S., & Ahmad. A. R., (2020). Stimulating organizational citizenship behavior by applying organizational commitment and satisfaction. *International Journal of Social Sciences and Economics Review, 2*(2), 20–27.

Meyer, J. P., & Allen. N. J., (1997). *Commitment in the Workplace: Theory, Research, and Application.* Sage.

Meyer, P. J., Stanley, D. J., Herscovitch, L., & Topolnytsky. L., (2002). Affective, continuance, and normative commitment to the organization: A meta-analysis of antecedents, correlates, and consequences. *Journal of Vocational Behavior, 61*, 20–52.

Mohamed, M. S., & Anisa. H., (2012). Relationship Between Organizational Commitment and Organizational Citizenship Behavior. *IUP Journal of Organizational Behavior, 11*(3), 7–22.

Morrison, E. W., (1996). Organizational citizenship behavior as a critical link between HRM practices and service quality. *Human Resource Management, 35*(4), 493–512.

Mosley, R. W., (2007). Customer experience, organizational culture, and the employer brand. *Journal of Brand Management, 15*(2), 123–134.

Ngo, L. V., Nguyen, N. P., Huynh, K. T., Gregory, G., & Cuong. P. H., (2019). Converting internal brand knowledge into employee performance. *Journal of Product & Brand Management.*

O'Reilly, C. A., & Chatman. J., (1986). Organizational commitment and psychological attachment: The effects of compliance, identification, and internalization on prosocial behavior. *Journal of Applied Psychology, 71*(3), 492–499.

Obedgiu, V., Bagire, V., & Mafabi. S., (2017). Examination of organizational commitment and organizational citizenship behavior among local government civil servants in Uganda. *Journal of Management Development, 36*(10), 1304–1316.

Organ, D. W., (1988). *Organizational Citizenship Behavior: The Good Soldier Syndrome*. Lexington Books/DC Heath and Com.

Organ, D., & Ryan. K., (1995). A meta-analytic review of attitudinal and dispositional predictors of organizational citizenship behavior. *Personnel Psychology, 48*, 775–808.
Paramaartha, D. N., Mukhtar, M., & Akbar. M., (2019). The effects of POS, interpersonal justice, and affective commitment on OCB of senior secondary teachers. In: *1st International Conference on Innovation in Education (ICoIE 2018)* (pp. 625–630). Atlantis Press.
Piercy, C. W., & Carr. C. T., (2020). Employer reviews may say as much about the employee as they do the employer: Online disclosures, organizational attachments, and unethical behavior. *Journal of Applied Communication Research, 48*(5), 577–597.
Podsakoff, P. M., MacKenzie, S. B., Moorman, R. H., & Fetter. R., (1990). Transformational leader behaviors and their effects on followers' trust in leader, satisfaction, and organizational citizenship behaviors. *The Leadership Quarterly, 1*(2), 107–142.
Podsakoff, P. M., MacKenzie, S. B., Paine, J. B., & Bachrach. D. G., (2000). Organizational citizenship behaviors: A critical review of the theoretical and empirical literature and suggestions for future research. *Journal of Management, 26*(3), 513–563.
Preacher, K. J., & Hayes. A. F., (2004). SPSS and SAS procedures for estimating indirect effects in simple mediation models. *Behavior Research Methods, Instruments, & Computers, 36*(4), 717–731.
Rafiq, M., & Ahmed. P. K., (2000). Advances in the internal marketing concept: Definition, synthesis, and extension. *Journal of Services Marketing, 14*(6), 449–462.
Reichheld, F. F., (2003). *The One Number You Need to Grow*. Harvard business review. https://hbr.org/2003/12/the-one-number-you-need-to-grow (accessed on 11 March 2022).
Rhoades, L., & Eisenberger. R., (2002). Perceived organizational support: A review of the literature. *Journal of Applied Psychology, 87*(4), 698–714.
Rhoades, L., Eisenberger, R., & Armeli. S., (2001). Affective commitment to the organization: The contribution of perceived organizational support. *Journal of Applied Psychology, 86*(5), 825–836.
Rogers, J. D., Clow, K. E., & Kash. T. J., (2014). Increasing job satisfaction of service personnel. *Journal of Service Marketing, 8*(1), 14–26.
Satuf, C., Monteiro, S., Pereira, H., Esgalhado, G., Afonso, R., & Loureiro. M., (2018). The protective effect of job satisfaction in health, happiness, well-being, and self-esteem. *International Journal of Occupational Safety and Ergonomics, 24*(2), 181–189.
Schneider, B., & Bowen. D. E., (1993). The service organization: Human resources management is crucial. *Organizational Dynamics, 21*(4), 39–52.
Schneider, B., (1987). The people make the place. *Personnel Psychology, 40*(3), 437–453.
Storbacka, K., Strandvik, T., & Grönroos. C., (1994). Managing customer relationships for profit: The dynamics of relationship quality. *International Journal of Service Industry Management, 5*(5), 21–38.
Sull, D., Sull, C., & Bersin. J., (2020). Five ways leaders can support remote work. *MIT Sloan Management Review, 61*(4), 1–10.
Sun, T., Ayoun, B., & Calhoun. J., (2013). The effect of organizational commitment on word-of-mouth intentions in recruitment in China. *Journal of Human Resources in Hospitality & Tourism, 12*(2), 192–216.

Terglav, K., Ruzzier, M. K., & Kaše. R., (2016). Internal branding process: Exploring the role of mediators in top management's leadership–commitment relationship. *International Journal of Hospitality Management, 54*, 1–11.
Towler, A., Lezotte, D. V., & Burke. M. J., (2011). The service climate-firm performance chain: The role of customer retention. *Human Resource Management, 50*(3), 391–406.
Van, N. L. T., Lu, V. N., Hill, S. R., & Conduit. J., (2019). The mediating role of brand knowledge on employees' brand citizenship behavior: Does organizational tenure matter?. *Australasian Marketing Journal (AMJ), 27*(3), 169–178.
White, J. G., (2008). *Examining Net Promoter Score (NPS) as An Effective Evaluation Tool for Measuring Customer and Employee Satisfaction (Order No. 3335328)*. Available from ProQuest One Academic. (304823545). Retrieved from: https://0-search-proquest-com.biblioteca-ils.tec.mx/docview/304823545?accountid=11643 (accessed on 18 February 2022).
World Health Organization (2020). WHO Director-General's Opening Remarks at the Media Briefing on COVID-19. https://www.who.int/director-general/speeches/detail/who-director-general-s-opening-remarks-at-the-media-briefing-on-covid-19---11-march-2020 (accessed on 11 March 2022).
Workplace, (n.d.). Virgin. *Opening New Horizons for Virgin with Smarter Technology and Teamwork.* https://www.workplace.com/case-studies/virgin (accessed on 18 February 2022).

CHAPTER 10

Adapting Marketing Channels During COVID-19

GLORIA CAMACHO

Marketing and Analysis Department, Tecnológico de Monterrey, Campus Monterrey, Monterrey, Mexico, E-mail: Camacho.Gloria@tec.mx

ABSTRACT

Before the pandemic of COVID-19, many companies considered being part of the digital world, but e-commerce did not seem urgent at that time. However, the crisis originated by this pandemic motivates firms to implement actions to face actual challenges, such as a drop in demand, disruptions in the supply chain, the close of physical stores, and changes in consumers' buying behavior. Thus, this chapter aims to explore the adaptation of marketing channels in a pandemic environment. The chapter presents a Mexican firm's case in the footwear industry. The firm strengthens its newly launched online store as an alternative to ensure its survival during its closed physical stores in the lockdown and the new normal by continuing with its relationship with its customers.

10.1 INTRODUCTION

Information technology, internet access, the new digital channels (i.e., mobile), and the growth of social media have changed consumers' buying behavior (Salmani, Partovi, and Banerjee, 2018) and retailing (Verhoef, Kannan, and Inman, 2015).

Marketing by Contingency in the Time of COVID-19: Overcoming Business Crises and Meeting Marketing Challenges. Alicia De La Peña, PhD & Juan Bernardo Amezcua Nunez, PhD (Eds.)

According to Online Sales Mexican Association, online sales have increased in the last years (Galeano, 2019). For example, from 2017 to 2018, the growth was 33%, and in 2019 digital retailing commerce increased by 34% (Euromonitor, 2020). Before the pandemic of COVID-19, many companies considered being part of the digital world. However, a crisis like the one caused by this pandemic, which a high degree of uncertainty and adverse outcomes, demands the implementation of actions to respond to the new challenges and ensure its survival (Fonseca and Acevedo, 2020). Therefore, COVID-19 has accelerated the adoption of digital transformation (World Footwear, 2020a) "digital tools have become a lifeline for local businesses previously reliant on in-store experiences, with technology allowing them to stay operational" (Coronado, 2020). For example, in Mexico, according to a study developed by the Online Sales Mexican Association in April 2020, consumers preferred to buy online because they did not want to leave their homes, and they wanted to avoid a lot of people during the lockdown. The most preferred categories to buy online were food delivery, fashion, supermarket, medicines, electronics (AMVO, n.d.). Thus, it became essential for firms to build an integrated digital presence (World Footwear, 2020a) to survive during the lockdown, and Mexican firms are not the exception.

This chapter is structured as follows: First, I present marketing channel strategies. Second, I show methodology and then the Flexi case. Afterward, I discuss how the chosen firm illustrates its marketing channels' adaptation in a pandemic context. Finally, I present a conclusion.

10.2 MARKETING CHANNELS

Initially, marketing channels focused on distributing system processes (Gundlach et al., 2006). With the evolution of marketing and its focus on delivering value to the end-user, marketing channel conception evolved. Thus, marketing channels are where the customer value-creating processes are the most pronounced in the supply chain (Kozlenkova et al., 2015, p. 587).

According to Palmatier et al. (2004, as cited in Kozlenkova et al., 2014), marketing channels are formed by the different organizations that help to put the product or service available for the consumers. Traditionally, they include manufacturers, intermediaries (i.e., retailers), and final consumers in B2B or B2C markets.

Firms use multiple channels (i.e., brick-and-mortar stores, online) to satisfy their consumers' needs (Kozlenkova et al., 2015; Neslin and Shankar, 2009; Müller-Lankenau, Wehmeyer, and Klein, 2006) and to increase their market shares (Salmani, Partovi, and Banerjee, 2018). Multichannel marketing "enables firms to build lasting customer relationships by simultaneously offering their customers and prospects information, products, services, and support (or any combination of these) through two or more synchronized channels" (Rangaswamy and Bruggen, 2005, p. 6). Under this approach, firms focus on optimizing each channel's performance separately (Cui et al., 2021).

Empirical evidence suggests that multichannel customers have more sales volume and are more valuable for a firm. Also, multichannel use is associated with higher customer value (Neslin and Shankar, 2009). However, the involved channels need to share information (Rangaswamy and Bruggen, 2005). Some of the limitations of multichannel marketing are the lack of integration between channels, narrow channel scope, lack of consideration of the impact of each channel independently and all of them together in customer experience (Payne, Peltier, and Barger, 2017), and siloed approach to managing the different channels (McCarthy Byrne, 2017).

With the introduction of new technologies, and the changes in consumer behavior, such as the adoption of mobile devices and the use of social media (Bherman and Thelen, 2018), omnichannel emerged. Omnichannel is understood as "synergetic management of the numerous available channels and customer touchpoints, in such a way that the customer experience across channels and the performance over channels is optimized" (Verhoef et al., 2015, p. 176).

Omnichannel marketing is defined as "the synergistic management of all customer touchpoints and channels both internal and external to the firm to ensure that the customer experience across channels as well as firm-side marketing activity, including marketing-mix and marketing communication (owned, paid, and earned), is optimized for both firms and their customers" (Cui et al., 2021, p. 104). It allows customers to use different combinations of channels and devices at each phase of their purchase journey (Bherman and Thelen, 2018). For example, a consumer can use a smartphone to look for a product, a laptop to get product information, a tablet to check last prices, make a credit card payment in an online store, and then arrange the return at a local store.

With the evolution from multichannel to omnichannel, it became necessary to integrate the different marketing channels (Piotrowicz and Cuthberson, 2014). This integration seeks to promote and deliver positive consumer experiences (Gundlach et al., 2006), to offer a seamless shopping experience to consumers (McCarthy Byrne, 2017), and to be competitive in the marketplace (Salmani, Partrovi, and Banerjee, 2018).

In omnichannel marketing, firms need to manage the store, online website, direct marketing, mobile channels (i.e., tablets, apps, smartphones), social media, and other customer touchpoints such as mass communication channels (i.e., TV, radio) (Verhoef et al., 2015). Also, consumers have unrestricted access to inventory across different channels (McCarthy Byrne, 2017). It is important to remark that consumers interact with the brand under the omnichannel approach, not with the channel (Savastano, Bellini, D'Ascenzo, and De Marco, 2019). Some characteristics of the omnichannel marketing are the use of uniform message across channels and devices, a seamless customer's experience no matter which channel or device is used, the understanding of consumer behavior, consumers do not want to make efforts to move from one channel to another (Bherman and Thelen, 2018; Cui et al., 2021).

10.3 METHODOLOGY

This chapter follows a case study research strategy (Eisenhardt, 1989). Secondary data support the present analysis (Godoy, 1995). I collected information between September 2020 and November 2020 from online sources such as newspapers, press releases, publications from different organizations, the firm's app, website, and social media, such as the Facebook page and the Instagram account. The case illustrates how a Mexican firm adapted its marketing channels during COVID-19. Flexi was chosen through purposeful sampling (Marshall, 1996), considering it is the second brand sold in Mexico's footwear sector (Expansión, 2017). Just a few months before COVID-19, it dabbled in digital retail through the launch of its internet shopping site (América Retail, 2019). The Flexi case exemplifies the cautious adoption of online channels to offer its products to its consumers before COVID-19 in some Mexican firms. This case provides insights into the consequences of COVID-19 in the digitalization of selling and the online channel management to continue operations during difficult times derived from the lockdown in Mexico during April and May 2020. To sum up, the Flexi case shows how this firm adapted its marketing channels to face COVID-19 impacts.

10.4 THE FLEXI CASE

10.4.1 MEXICO, COVID-19 AND FOOTWEAR INDUSTRY

The World Health Organization (WHO) declared COVID-19 a global pandemic on March 12, 2020 (Mexperience, 2020a). On March 24, according to the Innovation and Competitiveness Centre at the Chamber of the Footwear Industry of the State of Guanajuato, which is the most important footwear manufacturer in Mexico (i.e., in 2018, Guanajuato made over 180 million pairs of shoes) (Shoes from Mexico, 2017), footwear manufacturing plants were working with the implementation of sanitary measures (World Footwear, 2020b). By the end of March, Mexico declared a public health emergency and suspended all non-essential activities until June. The lockdown forced businesses to close (Mexperience, 2020b). The essential activities were "the provision of medical services and supplies, grocery delivery services, operation of grocery stores, restaurant delivery and carryout services, assurance of public safety, fundamental economic functions and government social program, as well as critical infrastructure, construction, and manufacturing of transportation equipment" (US Embassy Mexico City, 2020).

At the end of April, the Mexican footwear association and others from different Latin-American countries were concerned about stopping manufacturing, distribution, and commercialization of footwear because of the lockdown (World Footwear, 2020c). Mexico is the ninth footwear producer in the world. In Mexico, the footwear industry employs about 215,000 persons. Also, it produces around 250,000 million pairs of shoes (Verján, 2020). Flexi is the second brand sold in Mexico's footwear sector (Expansión, 2017).

In May, the Mexican government developed a traffic light system used by the different states to reopen activities on June 1, 2020, which goes from red to green. In the color red, only essential economic activities are open. In yellow, closed public spaces reopen with a reduced capacity, and in green all activities are open with a safety distance (KPMG, 2020).

On June 3, Guanajuato's footwear sector re-started operations under sanitary measures (World Footwear, 2020c), including manufacturing plants, such as those owned by Flexi.

10.4.2 COMPANY PROFILE

Grupo Flexi is an industrial group, which focuses on the design, manufacture, and commercialization of footwear. The headquarter is in León, Guanajuato, and its main distribution channel is its Flexi shoe shop chain (EMIS, 2020).

Flexi began operations in 1935 when Roberto Plascencia Gutiérrez founded it. It started with a small group of artisans. They manufactured shoes for children under the Cesar brand. Later, the brand changed its name to Duende. In 1965, when it started men's shoe production, the firm's name changed to Flexi (Somos Industria, 2009–2018). The new name would allow the identification of flexible and comfortable shoes offer by this group (Redacción AM, 2017).

In 2004, the group opened 30 Flexi shoe shops in Mexico, and in 2014, there were around 300 shoe shops. Flexi opened its first distribution center in 2009 and its first shop in the United States in 2012. In 2014, Flexi launched its first loyalty program for its loyal customers (Flexi, 2019). In Mexico, Flexi sells its shoes in departmental stores such as Palacio de Hierro, Liverpool, and Coppel (Expansión, 2017). Flexi has an international presence in the United States, Japan, Spain, and Latinamerican countries (Ijelman, 2018).

Flexi's mission is "At Flexi Group, a Mexican firm focused on manufacture and distribution of footwear and accessories, we seek to transcend as a leading brand, which fulfills consumers, clients, and suppliers' expectations, and equitable distributing the value generated among shareholders, employees, government, and society" (Flexi, 2019).

10.4.3 FLEXI AND ITS ONLINE CHANNEL

10.4.3.1 BEFORE COVID-19

10.4.3.1.1 Flexi's App

Currently, with mobile computing and shopping apps, consumers have much greater visibility into the stores' inventory, pricing, and promotional activity (McCarthy Byrne, 2017, p. 6). According to a study developed by Deloitte, 59% of the participants preferred the smartphone to buy online (Deloitte, 2021). Also, mobile app users tend to be more loyal and engaged (Sailthru, 2021, p. 6).

Mobile apps have contributed to the rise of mobile shopping. The perceived ease of use is one of the app attributes that impact consumers' intention to make purchases with the app (Newman, Wachter, and White, 2018).

In 2013, Flexi launched its app, which was available on smartphones and tablets (Flexi, 2019). If you visit social media, such as Flexi's Facebook page, there is a post on September 19, 2018, promoting the new experience with Flexi App and inviting consumers to download it. The message also mentions that the app is available for Android and iOS (Post Flexi, 2018). On November 22, 2020, the app includes the collection and trends of the autumn-winter 2018 season. The outdated collection in the app shows that it does not really fulfilling customers' need for actual information. In the initial screen, consumers can look for stores; this section links to the actual online store's website that consumers will check from their mobile. On the same screen, there is a bulletin bottom that does not work. A contact section sends you to the different customer service channels, such as a telephone number, a chat available from Monday to Friday (8 am to 5 pm), e-mail, and frequently asks questions (FAQS). The main issue with the Flexi's app is that it is not working, and it has outdated information.

10.4.3.1.2 Flexi's Online Store

Since 2015, Flexi recognized the importance of being part of the digital world and offering its consumers the opportunity to acquire their shoes online and pick-up them in a physical store or receive them at home (Imagen, 2015). Flexi identified that some benefits of having its online store are that brand followers and consumers will find the entire collection offered by Flexi each season (Mares, 2019). Even though Flexi has a presence in e-commerce platforms, such as Amazon, it launched its own online shopping store in the Mexican market (América Retail, 2020). According to Jorge Rosas, Flexi's Marketing Director, Flexi wanted to offer a competitive service, so it started its online store, which has the entire collection and discounts (Ochoa, 2020).

10.4.3.2 COVID-19 AND THE NEW NORMAL

The lockdown forced non-essential Mexican businesses to close (Mexperience, 2020b). Thus, Flexi used its website to communicate the close of all of its Flexi shoe shops from April 2020 until new notice. The group remarked that its priority was the health and safety of its clients and its employees. The group promoted its online store https://www.flexi.com.mx, offered free shipping with its logistical partners (i.e., DHL, and Estafeta), and gave information regarding customer service which includes two additional telephone lines (Flexi, 2020).

Before COVID-19, e-commerce seemed to be irrelevant for Flexi operations. However, Rosas (Flexi's Marketing Director) found its online store helped move the firm's finances during the lockdown in April and May 2020 (Ochoa, 2020). In April-May 2020, Flexi posted the link of its web site in its social media, such as Facebook. On the one side, based on some of consumers' comments, they liked the shoes presented by Flexi and they mentioned they already bought them, and, in some cases, they posted their own pictures using Flexi's shoes or pictures when they received their package with their Flexi's shoes. On the other side, some customers mentioned that they wanted to acquire Flexi's shoes when their shoe shop open again.

10.4.3.2.1 Flexi's Website

Consumer behavior is changing. They want to optimize searching time and want a unique experience in their purchase journey. In online context, consumers evaluate products through design elements (i.e., web site) (Cachero-Martínez and Vázquez-Casielles, 2021; Bleier, Harmeling, and Palmatier, 2019). Thus, retailers need to offer a shopping experience through aspects that they can control such as web design (Cachero-Martínez and Vázquez-Casielles, 2021).

Online customer experience is understood as "the extent to which a website provides consumers with resourceful and helpful information" (Lim and Ting, 2012, p. 51 as cited in Bleier, Harmeling, and Palmatier, 2019, p. 99). According to Bleier, Harmeling, and Palmatier (2019), there are different elements that should be considered in website design to create an online customer experience, such as descriptive product detail, bulleted product features, return policy information, zoom of product features that would not be visible, people using the product, the size of the pictures used in the website, product video, customer's reviews, expert endorsements, and comparison matrix.

The main page includes Flexi's logo, find your store menu, register, and begin session section, account, favorites, cart, and menus for ladies, gentlemen, kids, Flexi country, and sales. Also, there is a statement that says that there is free shipping in all shopping.

In the middle of the website, the visitor will find the most sold footwear, including their prices. There are small icons that focus on: Free shipping, safety shopping, return policy, and support.

At the bottom, consumers have different menus, which include: (i) Us (i.e., History, mission, values; competitive advantage, quality, attributes, footwear construction, and job bank). (ii) Help (i.e., contact, FAQS, payment verification, size's guide, promotion terms, and terms and conditions). (iii) Client (i.e., medium wholesale, invoice, international sales (in English), and accessories). There is a space titled Buy safety with a statement of a site 100%, a Trust seal from Internet Association, and the Online Sales Mexican Association's logo.

Finally, the main page includes the different social media in which Flexi is present, such as Facebook, YouTube, Instagram, Pinterest, and Twitter. Flexi created its Facebook page in 2012, and it has 2,671,864 followers (Flexi, n.d.a). On Instagram, Flexi has about 1, 112 publications and 185,000 followers by November 22, 2020, and it offers a link to its website (Flexi, n.d.b). Flexi joined Twitter in 2012, and it based on tables elements, Flexi website fulfills most of them to create an online customer experience. However, it would improve through the addition of lifestyle pictures, comparison matrix and recommendation agent.

10.5 DISCUSSION

The World Footwear's experts estimate a decrease of 20–30% in footwear consumption worldwide (APICCAPS, 2020a). Also, changes in consumers' buying behavior include purchasing power, price, and fashion trends as the most critical drivers for buying shoes (World Footwear, 2020e). Furthermore, consumers will pay 3% more for online shopping and home delivery, 4% more for national production shoes or from recycled materials, 8% more for extended durability, and 9% more for customization (APICCAPS, 2020a).

Given this context, it seems like the pandemic environment offers the footwear industry an opportunity to adopt technology to deal with the drop in demand and the changing preferences in consumption, such as an increased preference for online shopping. In the specific case of Flexi, six months before the pandemic, it launched its online store. However, the physical shoe stores were the primary distribution channel for the footwear (América Retail, 2019).

Flexi launched its online channel, and based on its posts on Facebook, the firm uses this social media to guide its consumers to the online site, as Müller-Lankenau, Wehmeyer, and Klein (2006) suggest. After the pandemic with Mexico's lockdown, it is possible to say that Flexi used its online store as a lifeguard to survive the closed of its physical shoe shops, considering that they are not essential activities according to the Mexican government. It is important to remark that this situation with the online channel of firms which are part of the footwear industry worldwide contributed to maintaining and, in some cases, to increase revenues during 2020. For example, CCC, a Polish group, reported a revenue increase in the second and third quarter of 2020 with strong sales in its e-commerce channel, representing about 95% of its sales (World Footwear, 2020f). Adidas reported a decline in its physical stores and a 93% increase in sales in its online channel (World Footwear, 2020g), and Wolverine, the Michigan group, increased 90% in sales through its e-commerce during the first months of the second quarter of 2020 (World Footwear, 2020h).

Academics suggest an online channel will increase a firm's market share (Salmani, Partovi, and Banerjee, 2018). A panel of experts agreed that footwear industry online stores (i.e., general sellers, multi-brand online shoes, own-brand online retail stores) would increase their market share. Physical stores will reduce it (APICCAPS, 2020a). Therefore, retailers should redefine their physical stores and they need to use technology to integrate their channels and engage customers (Savastano, Bellini, D'Ascenzo and Marco, 2019). Thus, no matter the industry, firms need to be part of e-commerce if they want to survive in post-covid arena.

After analyzing the case, one question emerged: What is the next for Flexi and the managing of its multiple marketing channels? I would say that it is continue working in the integration of its marketing channels (Piotrowicz and Cuthberson, 2014) and going beyond through an omnichannel model (Verhoef et al., 2015) as an attempt to offer the same experience to the customer no matter which marketing channel they choose to acquire Flexi's products. This is marketing implication for this chapter, the challenge is to allow consumers to move from online to physical channels as part of an omni-channel approach to brand engagement (Payne, Peltier, and Barger, 2017).

10.6 CONCLUSION

Digital presence covered relevance after the shift from store to online shopping with COVID-19. It is crucial to keep loyalty, allow customers to connect with the brand, and maintain customer engagement. Besides, it is useful to keep the updated operation of stores on the firm's website; be clear about the instructions regarding product delivery and your policy for exchange and return (APPICAPS, 2020b).

As a result of the pandemic crisis, Grupo Flexi used technology and innovation to face the "new normal" challenges. Technology became one of the elements that promote continuity and recovery of this firm through the strength of the customer experience in its online store (SAP Noticias, 2020). Now, Flexi needs to work on its omnichannel strategy to offer complete integration of its channels and the same experience if a consumer buys in the online store, in the physical shoe shop, or through the app.

KEYWORDS

- **COVID-19**
- **endorsements**
- **Flexi website**
- **Online Sales Mexican Association**
- **online shopping**

REFERENCES

América Retail, (2019). *México Flexi Acelera Hasta 500 Tiendas Este año y Lanza su Plataforma de E-Commerce.* América Retail. https://www.america-retail.com/mexico/mexico-flexi-acelera-hasta-500-tiendas-este-ano-y-lanza-su-plataforma-de-ecommerce/ (accessed on 18 February 2022).

AMVO, (n.d.) *Report 2- Impacto COVID-19 en Venta Online Mexico.* AMVO. https://www.amvo.org.mx/estudios/reporte-2-impacto-covid-19-en-venta-online-mexico/ (accessed on 18 February 2022).

APICCAPS, (2020a). *World Footwear Business Conditions Survey,* 1–16. https://www.worldfootwear.com/news/impact-of-the-pandemic-on-footwear-consumption/5735.html?social_network=google&new_user=no (accessed on 18 February 2022).

APPICAPS, (2020b). *World Footwear Guide for Reopening Retail.* https://www.worldfootwear.com/news/is-your-store-or-brand-ready-for-customers-after-covid-19/5301.html (accessed on 18 February 2022).

Bherman, B., & Thelen. S., (2018). Planning and implementing an effective omnichannel marketing program. *International Journal of Retail & Distribution Management, 46*(7), 598–614.

Bleier, A., Harmeling, C. M., & Palmatier. R. W., (2019). Creating effective online customer experiences. *Journal of Marketing, 83*(2), 98–119.

Cachero-Martínez, S., & Vázquez-Casielles. R., (2021). Building consumer loyalty through e-shopping experiences: The mediating role of emotions. *Journal of Retailing and Consumer Services, 60*, 2–10.

Coronado, M., (2020). *Digitalization, an Enabler of Purpose.* https://0-www-portal-euromonitor-com.biblioteca-ils.tec.mx/portal/analysis/tab (accessed on 18 February 2022).

Cui, T. H., Ghose, A., Halaburda, H., Iyengar, R., Pauwels, K., Sriram, S., Tucker, C., & Venkataraman. S., (2021). Informational challenges in omnichannel marketing: Remedies and future research. *Journal of Marketing, 85*(1), 103–120.

Deloitte, (2021). *Digital Consumer Trends en Mexico 2020.* Deloitte. https://www2.deloitte.com/content/dam/Deloitte/mx/Documents/technology/2021/DigitalCT_2020.pdf (accessed on 18 February 2022).

Eisenhardt, K. M., (1989). Building theories from case study research. *Academy of Management Review, 14*(4), 532–550.

EMIS, (2020). *Grupo Flexi, S. A. De C. V. (México).* https://www.emis.com/php/company-profile/MX/Grupo_Flexi_SA_de_CV_es_2039931.html (accessed on 18 February 2022).

Euromonitor, (2020). *Economies and Consumer Annual Data- Mexico.* https://0-www-portal-euromonitor-com.biblioteca-ils.tec.mx/portal/StatisticsEvolution/index# (accessed on 18 February 2022).

Expansión, (2017). *Roberto Plascencia Saldaña: El Empresario que Convirtió a Flexi en Multinacional.* Expansión. https://expansion.mx/empresas/2017/10/16/roberto-plasencia-saldana-el-empresario-que-convirtio-a-flexi-en-multinacional (accessed on 18 February 2022).

Flexi [@flexi], (n.d.a). *Flexi [Facebook Profile].* Facebook. Retrieved from: https://www.facebook.com/flexi/ (accessed on 18 February 2022).
Flexi [@flexi], (n.d.c). *Flexi [Twitter Profile].* Twitter. Retrieved from: https://www.instagram.com/goflexi/ (accessed on 18 February 2022).
Flexi [@goflexi], (n.d.b). *Flexi [Instagram Profile].* Instagram. Retrieved from: https://www.instagram.com/goflexi/ (accessed on 18 February 2022).
Flexi, (2019). *Nuestra Historia.* https://somos.flexi.com.mx/historia/ (accessed on 18 February 2022).
Flexi, (2020). #CómodoEnCasa. https://somos.flexi.com.mx/comodo-en-casa/?utm_source=newsletter&utm_medium=email&utm_term=tiendas&utm_campaign=covid (accessed on 18 February 2022).
Fonseca, L. M., & Azevedo. A. L., (2020). COVID-19: Outcomes for global supply chains. *Management & Marketing. Challenges for the Knowledge Society, 15*, 424–438.
Galeano, S., (2019). *Las Ventas Onine en México se Incrementan 59% Durante Primer Semestre de 2019.* Marketing 4 ecommerce MX. https://marketing4ecommerce.mx/las-ventas-online-en-mexico-se-incrementan-59-durante-primer-semestre-de-2019/ (accessed on 18 February 2022).
Godoy, A. S., (1995). Qualitative Research: Fundamental Types. *Business Administration Journal, 35*(3), 20–29.
Gundlach, G. T., Bolumole, Y. A., Eltantawy, R. A., & Frankel. R., (2006). The changing landscape of supply chain management, marketing channels of distribution, logistics, and purchasing. *Journal of Business & Industrial Marketing, 21*(7), 428–438.
Ijelman, C., (2018). *Datos y Cifras de Marcas de Moda en México: Flexi.* Fashion United. https://fashionunited.mx/noticias/retail/datos-y-cifras-de-marcas-de-moda-de-mexico-flexi/2018090525146 (accessed on 18 February 2022).
Imagen, (2015). *Has Usado Zapatos Pero ¿Conoces la Historia de Flexi?* Dinero en Imagen. https://www.dineroenimagen.com/2015-11-27/65194 (accessed on 18 February 2022).
Kozlenkova, I. V., Hult, G. T. M., Lund, D. J., Mena, J. A., & Kekec. P., (2015). The role of marketing channels in supply chain management. *Journal of Retailing, 91*(4), 586–609.
KPMG (2020). *Mexico Government and Institution Measures in Response to COVID-19.* https://home.kpmg/xx/en/home/insights/2020/04/mexico-government-and-institution-measures-in-response-to-covid.html (accessed on 18 February 2022).
Mares, A., (2019). *Flexi Lanzará su Canal de E-commerce Este 2019.* Fashion Network. https://mx.fashionnetwork.com/news/Flexi-lanzara-su-canal-de-e-commerce-este-2019,1068242.html (accessed on 18 February 2022).
Marshall, M. N., (1996). Sampling for qualitative research. *Family Practice, 13*(6), 522–525.
McCarthy, B. T. M., (2017). Omnichannel: How will it impact retail forecasting and planning processes? *Journal of Business Forecasting, 35*(4), 4–9.
Mexperience, (2020a). *Mexico and the Coronavirus COVID-19* (Updated). https://www.mexperience.com/mexico-and-the-corona-virus-covid-19/ (accessed on 18 February 2022).
Mexperience, (2020b). *Your Local Mexican Economy Needs Your Support Now.* https://www.mexperience.com/step-up-and-support-your-local-mexican-economy/ (accessed on 18 February 2022).

Müller-Lankenau, C., Klein, S., & Wehmeyer. K., (2004). Developing a framework of multichannel strategies -an analysis of cases from the grocery retail industry. *Paper Presented at 17th Bled Electronic Commerce Conference e-Global.* https://aisel.aisnet.org/cgi/viewcontent.cgi?article=1060&context=bled2004 (accessed on 18 February 2022).

Müller-Lankenau, C., Wehmeyer, K., & Klein. S., (2006). Strategic channel alignment: An analysis of the configuration of physical and virtual marketing channels. *Information Systems and e-Business Management, 4*(2), 187–216.

Neslin, S. A., & Shankar. V., (2009). Key issues in multichannel customer management: Current knowledge and future directions. *Journal of Interactive Marketing, 23*, 70–81.

Newman, C. L., Wachter, K., & White. A., (2018). Bricks or clicks? Understanding consumer usage of retail mobile apps. *Journal of Services Marketing, 32*(2), 211–222.

Ochoa, C., (2020). *Por Cierre de Tiendas Ante Coronavirus, Flexi Alcanza Meta de Usuarios Digitales*. Milenio. https://www.milenio.com/negocios/coronavirus-flexi-alcanza-meta-compradores-digitales (accessed on 18 February 2022).

Payne, E. M., Peltier, J. W., & Barger. V. A., (2017). Omni-channel marketing, integrated marketing communications and consumer engagement. A research agenda. *Journal of Research in Interactive Marketing, 11*(2), 185–197.

Piotrowicz, W., & Cuthbertson. R., (2014). Introduction to the special issue information technology in retail: Toward omnichannel retailing [Abstract]. *International Journal of Electronic Commerce, 18*(4), 5–16.

Post Flexi, (2018). *Enjoy the New Flexi App Experience* [Image attached]. Facebook. https://www.facebook.com/flexi/posts/disfruta-la-nueva-experiencia-de-la-app-flexi-y-descubre-las-tendencias-que-marc/1812109278908796/ (accessed on 18 February 2022).

Rangaswamy, A., & Van, B. G. H., (2005). Opportunities and challenges in multichannel marketing: An introduction to the special issue. *Journal of Interactive Marketing, 19*(2), 5–11.

Redacción, AM, (2017). *They Forge a Leading Company*. AM. https://www.am.com.mx/noticias/Forjan-empresa-lider-20171016-0065.html (accessed on 18 February 2022).

Sailthru, (2021). *The 4 Curbside Customers to Know in 2021.* Sailthru https://www.sailthru.com/resources/bopisandcurbsidepickup/ (accessed on 18 February 2022).

Salmani, Y., Partovi, F., & Banerjee. A., (2018). Customer-driven investment decisions in existing multiple sales channels: A downstream supply chain analysis. *International Journal of Production Economics, 204*, 44–58.

SAP Noticias, (2020). *Empresas Mexicanas Destacan con su Innovación en Foro Internacional.* https://news.sap.com/latinamerica/2020/06/empresas-mexicanas-destacan-con-su-innovacion-en-foro-internacional/ (accessed on 18 February 2022).

Savastano, M., Bellini, F., D'Ascenzo, F., & De Marco, F., (2019). Technology adoption for the integration of online-offline purchasing Omnichannel strategies in retail environment. *International Journal of Retail & Distribution Management, 47*(5), 474–492.

Shoes from Mexico, (2017). *Footwear from Guanajuato and Internationalization.* Shoes from Mexico. https://shoesfrommexico.com/footwear-from-guanajuato-and-internationalization/ (accessed on 18 February 2022).

Somos Industria, (2009–2018). *Consolida Grupo Flexi su Crecimiento.* Somos Industria. https://www.somosindustria.com/articulo/consolida-grupo-flexi-su-crecimiento/ (accessed on 18 February 2022).

US Embassy Mexico City, (2020). *Health Alert: Mexico COVID-19 Update.* https://mx.usembassy.gov/health-alert-mexico-covid-19-08-12-2020/ (accessed on 18 February 2022).

Verhoef, P. C., Kannan, P. K., & Inman. J. J., (2015). From multi-channel retailing to omni-channel retailing introduction to the special issue on multi-channel retailing. *Journal of Retailing, 2*, 174–181.

Verján, R. M., (2020). *Infografía: Panorama de la Industria del Calzado en México.* Grupo Mundo Ejecutivo. https://mundoejecutivo.com.mx/exportadoras/2020/10/05/infografia-panorama-de-la-industria-del-calzado-en-mexico/ or https://www.facebook.com/MundoEjecutivoWeb/posts/3880398011990278/ (accessed on 18 February 2022).

Wilkinson, I. F., (1990). Toward a theory of structural change and evolution in marketing channels. *Journal of Macromarketing, 8*, 18–46.

World Footwear, (2020a). *Building a Digital Presence: What Steps to Consider?* https://www.worldfootwear.com/news/building-a-digital-presence-what-steps-to-consider/5698.html?social_network=google&new_user=no (accessed on 18 February 2022).

World Footwear, (2020b). *Mexico: Early Stages of COVID-19.* https://www.worldfootwear.com/news/mexico-early-stages-of-covid-19/4644.html (accessed on 18 February 2022).

World Footwear, (2020c). *Latin America: Footwear Associations Raise Their Concerns.* https://www.worldfootwear.com/news/latin-america-footwear-associations-raise-their-concerns/4856.html (accessed on 18 February 2022).

World Footwear, (2020d). *Industrial Production Re-starts in Guanajuato.* https://www.worldfootwear.com/news/industrial-production-re-starts-in-guanajuato/5045.html (accessed on 18 February 2022).

World Footwear, (2020e). *What Drives Sales of Footwear*? https://www.worldfootwear.com/news/what-drives-sales-of-footwear/5744.html?social_network=google&new_user=yes (accessed on 18 February 2022).

World Footwear, (2020f). *CCC with Fast Sales Recovery of the Online Channel.* https://www.worldfootwear.com/news/ccc-with-fast-sales-recovery-of-the-online-channel-/5607.html (accessed on 18 February 2022).

World Footwear, (2020g). *Adidas Nearly Doubles E-commerce Sales.* https://www.worldfootwear.com/news/adidas-nearly-doubles-e-commerce-sales-/5378.html (accessed on 18 February 2022).

World Footwear, (2020h). *Online Sales Deliver More than 40% of Wolverine's Revenue.* https://www.worldfootwear.com/news/online-sales-deliver-more-than-40-of-wolverines-revenue/5066.html (accessed on 18 February 2022).

CHAPTER 11

Omnichannel Strategies in Retail Places: How Consumers Changed Their Shopping Routines During the Pandemic

ENA YURITZE BARÓN LÓPEZ

Professor of the Department of Marketing of the Central University, Bogotá, Colombia, E-mail: ebaronl@ucentral.edu.co

ABSTRACT

The new sceneries for the brands change; they must contact the consumer by physical and digital channels. Around 10 years ago, the multichannel context began, and at this moment, the omnichannel is the challenge for retails. The purpose of omnichannel is that the channels are interchangeable and offer the same shopping experience. This condition makes it impossible to control the use of tracks by the firms. (Verhoef, Kannan, and Inman, 2015). Moreover, with the COVID-19 pandemic, brands need to make the most significant number of purchase possibilities available because consumers' willingness to purchase through digital channels has accelerated while buying in stores and supermarkets.

This chapter will explore the new consumer habits after the pandemic, specifically into the retail brands' channels and economics budgets. If the brand is small o big, it is essential to know the better option to give their clients and work to improve the experience and satisfy the necessity. This research focuses on the retail sector and the perception of brands and aims to know what changes have been in purchasing behavior. This

Marketing by Contingency in the Time of COVID-19: Overcoming Business Crises and Meeting Marketing Challenges. Alicia De La Peña, PhD & Juan Bernardo Amezcua Nunez, PhD (Eds.)

research focuses on the retail sector and the perception of brands and aims to know what changes have been in Buying behavior result by COVID-19 pandemic. The study was developed in the city of Bogotá-Colombia; 361 people responded to the survey. The main retail brands of Bogotá-Colombia were considered, which had different channels Online and Offline available for the consumer.

11.1 INTRODUCTION

Omnichannel is integration for all channels offered to the consumer. Omnichannel focuses on synergy and integration of available channels to streamline customer interactions and retailer control across channels (Li et al., 2018). One of the difficulties to organizations in this pandemic time was to take control of the deliverance of their products and services. Now it is more critical than other times to help the consumer to feel cared for and secure. From a marketing perspective, most brands will need to rip up their marketing playbook and drastically update their customer personalities and communication strategy. Consumers are adaptive and relative decision-makers (Babu and Kumar, 2020). Therefore, it is essential to establish the brands of retailers' brands and consumers' preference of the brands of the different commercial surfaces.

11.1.1 WHAT OMNICHANNEL MEANS?

Since consumers could navigate and interact with brands both in stores and on phones, tablets, laptops, retailers, banks, and services, they have had to adapt their marketing strategies to these new forms of consumption. Omnichannel widens the reach of points of sale by integrating the interaction between the consumer, the brand, and the channels. This results from the understanding of the omnichannel phenomenon (Verhoef, Kannan, and Inman, 2015).

Omnichannel essentially requires technological processes, and the challenge is the optimal coordination between technological processes and supply channels. (Saghiri, Wilding, Mena, and Bourlakis, 2017). Omnichannel buyers seek to find out, investigate, compare, and learn from other comparators from the same electronic device. Consumers demand to have the same purchasing conditions in all channels, which implies efficient distribution and adequate communication. Inevitably omnichannel embraces the idea of using multiple channels focused on integrating activities within and through these in line with the consumer's purchasing needs (Figure 11.1) (Ailawadi and Farris, 2017).

FIGURE 11.1 Omnichannel environment.

Source: Elaborated by the author.

Since the consumer has had the possibility of connecting simultaneously, sharing their experience with other consumers, the brands should know their needs since they can freely give their opinions on social networks about the strengths and weaknesses of experiences with brands. Within omnichannel retailing, the concept of brand experience is particular as it includes interactive channels such as social media and review sites. To complete even a single purchase, customers nowadays are using multiple channels. Customers are trying out products in a physical store but at the same time ordering them online using smartphones after comparing prices and checking reviews (Akter et al., 2019). As a result, the spaces for consumer interaction with the brands increased, and gradually the barriers became blurred.

Since the consumer could connect simultaneously, sharing their experience with other consumers, brands must know their needs since they can freely comment on the strengths and weaknesses of experiences with brands on social media. The omnichannel experience is crucial and occurs in interactive channels such as social networks and opinion and rating spaces. Even in a single purchase, customers today make use of the available channels, make use of the physical store, and at the same time the web pages and telephones; compare prices and verify (Akter et al., 2019). As a result, consumer interaction with the brands increased, and the barriers gradually blurred.

11.1.2 RETAIL BRANDS IN THE OMNICHANNEL ENVIRONMENT

For brands, the immediate crisis response was to shrink from public view. A primary principle of brand reputation is associative: brand owners assume that consumers will associate the brand's meaning and value(s) with whatever appears in proximity to it (Aronczyk, 2020); brands have had to review the consequences of consumer perception on their actions during the pandemic, particularly considering that the enterprise's image is reflected in the brand's actions. It has been shown that the images of the brands that the retailer distributes are affected by the distributor's image, and the image of the distributor is affected by the image of the brands that it distributes (Ailawadi and Keller, 2004). The assortment strategies influence the perception of the brands that the retailer welcomes. The perception of a brand is moderated by the merchandising or category strategies that the store implements. Hence managing the symbiosis between the retailer's brand and the brands it distributes is a very complicated task. It must be considered that the perception of these brands is linked to their communication strategies, product development, and prices. Consequently, retailers must regularly assess how their actions affect the sales performance and the brand experience simultaneously and steer clear short-term performance improvements that can negatively affect the brand experience (Blom, Lange, and Hess Jr, 2017). It could be considered that the differentiation concerning the brands of manufacturers and distributors, private or own, have almost been blurred because consumer strategies have exceeded the physical and production limits of the brands. Hence, the definitions made about prices, qualities, and characteristics between one and the other have evolved. The emergence of digital channels and new channels transformed the retail industry in recent years over the next decade. The integration of channels has driven the transformation with the aim to offer a unique experience, which is known as omnichannel (von Briel, 2018). Especially at this time, those who manage organizations seek to understand how to resolve consumer requests and maintain stability in the various channels, the brand. However, it is a very vulnerable element to that instability. It is also essential in generating confidence in the compliance of value promises; managers are certainly concerned about managing their brands across channels (Neslin et al., 2014). Most of the purchases are for personal care, food, and health products.

Branding choices also involve factors such as “proximity shopping,” which is a growing trend. The digital shopping culture that has been driven by the pandemic, attracting new consumers online, is a growing behavior that will continue after the outbreak (Forbes, 2020).

11.1.3 CONSUMER BEHAVIOR IN THE COVID-19 PANDEMIC

In the omnichannel context, studies on consumer behavior are the most advanced. The existing research on consumer behavior in the multichannel environment and, more recently, the omnichannel environment is focused on two main questions. On the one hand, the literature on the determinants of channel choice now focuses on the instances and sequences of moments in the customer’s walk (the contact points, according to Verhoef, Kannan, and Inman, 2015). But, on the other hand, a second investigation focuses on the profile and segmentation of consumers (Barwitz and Maas, 2018). Behavioral studies are intended to shed light on how the buying experience can be improved. A real challenge for organizations is providing a satisfying search experience for customers on different mobile devices for as long as the search takes. For the omnichannel customer, it is very frustrating to look for a product only to enter the store with a purchase decision and face that the price is different or that they even have other options (Cook, 2014). However, consumer behaviors changed because of the new living conditions due to the COVID-19 pandemic. The economic consequences due to the COVID-19 outbreak have affected all countries in the world.

This does not only affect the economy, but also the entire society, which has forced companies to make changes and consumers as well (Donthu and Gustafsson, 2020). These changes altered not only the economies but also the budget for specific categories of products and specifically in the way in which products and services are consumed. In this way, it could be considered that digital culture is reinforced to allow social and work meetings through screens, and the need for distancing accelerates the learning of the use of technologies to avoid physical encounters with others.

Some consumer trends focused on specific categories that were deemed necessary within consumer purchases in different countries. Many consumers globally continue to see their incomes fall, and optimism in an economic recovery has not seen a resurgence. Consequently, consumers are more aware of their expenses and consumption, since they expect the impact of COVID-19 to last more than four months. Consumers spend on essential products such as food and household products and decrease on non-essential products (Arora et al, 2020). This new situation opens up a wide range of possibilities in research on distribution and sales channels and consumer behavior.

11.2 RESEARCH QUESTION

This research focuses on the retail sector and the perception of brands and aims to know the changes in purchasing behavior derived from a pandemic?

11.3 METHODOLOGY

The study was developed in the city of Bogotá-Colombia. An instrument was designed and reviewed by colleagues with expertise in survey design. Around 15 people participated in the questionnaire's pre-tests to verify the instrument's understanding, response time, and supply ease. After receiving the options, the first adjustments were made to finally launch it by a specialized survey delivery platform. The sample consisted of people who had experience buying products in retail stores. Therefore, the sample is not probabilistic, and it was sent to many people who replicated the request; it was also exposed to various social networks. The study was developed in the city of Bogotá-Colombia; 361 were surveyed.

A survey was launched digitally through a specialized platform between April and May 2020. The survey's purpose was to collect data on demographics and consumer behavior before and after the pandemic. The months of April and May were especially sensitive times because consumers were learning to adapt to the new confinement conditions decreed by the city government (Bogotá). Therefore, they were developing new habits of life around them: consumption, social, and work habits between others (Table 11.1).

TABLE 11.1 Respondent Characteristics

Sex	Male 43.2%; Female 56.8%
Age	18–24; 21.3%
	25–34; 31.3%
	35–44; 30.2%
	45–54; 12.2%
	More than 55; 5%
Household Income	REF 242 Dollars per month (statutory minimum wage 2020 – Colombia)
	US$ 242< 2.2%
	1–3; 44.0%
	3–5; 22.6%
	5–7; 15.9%
	7–9; 9.5%
	9 > US$ 242; 5.8%
Education	Elementary education: 0.0%
	Secondary education: 17.5%
	Post-Secondary Educ-tech inst.: 18.6%
	Post Baccalaureate professional: 29.6%
	Post Baccalaureate master degree: 30.5%
	Other: 3.8%
Geographic Region	Colombia. City Bogotá
Occupation	Employee: 61%
	Unemployed: 3%
	Pensioner: 1%
	Independent: 14%
	Student: 21%

Source: Elaborated by the author.

11.3.1 TYPE OF QUESTIONS AND SCALES

The design of the questionnaire consisted of questions between dichotomous and Likert-type scales, and they focused on the following topics:

- The most frequent place of purchase of products and food;
- Change in spending on the purchase of products;
- Changes in the frequency of purchase of products;
- Most used retail brands.

11.4 RESULTS

We wanted to know the preference of brands to purchase and which retail format corresponded to the brands. The bar chart presented in Figure 11.2 shows the retail format of brands that prefer.

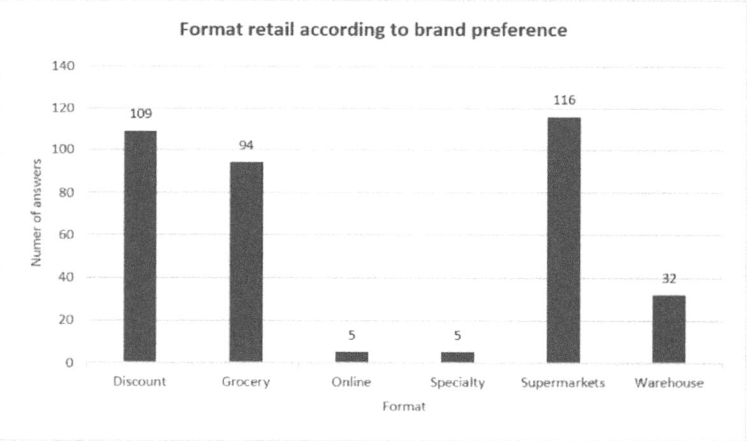

FIGURE 11.2 Format retail-brand preference.

Source: Elaborated by the author.

Supermarket brands have a 32% preference, 30% confirmed the choice of discounters' brands, and 26% prefer grocery stores (their owner personalizes the brands of these establishments). Brands of online and specialized formats have a 1% preference, respectively, and warehouse stores a

9%. On the other hand, those surveyed were asked about the changes in the purchase channel's use before and after the pandemic (Figure 11.3).

Do you normally buy your grocery for which of the following means?

In which of the following means or places did the purchase increase?

Options:

- On the website;
- In the mobile application;
- By phone;
- In the physical store.

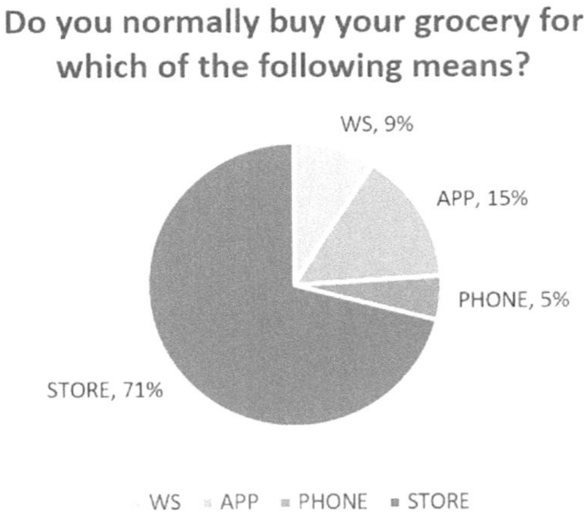

FIGURE 11.3 Frequent places for groceries and household products before a pandemic.

Source: Elaborated by the author.

Around 71% of people bought their products in the physical store, 15% through the mobile application, 9% through the web page, and 5% by phone call. The people prefer to go to the store or call traditional food and grocery stores to stock food and household products. About the question: In which of the following means or places did the purchase increase? The answers are shown in Figure 11.4.

Total responses (66%) correspond to the increase in purchases in the store, 10% increase in web pages, 15% in mobile applications, and 9% by telephones; interestingly, purchases by telephone increased.

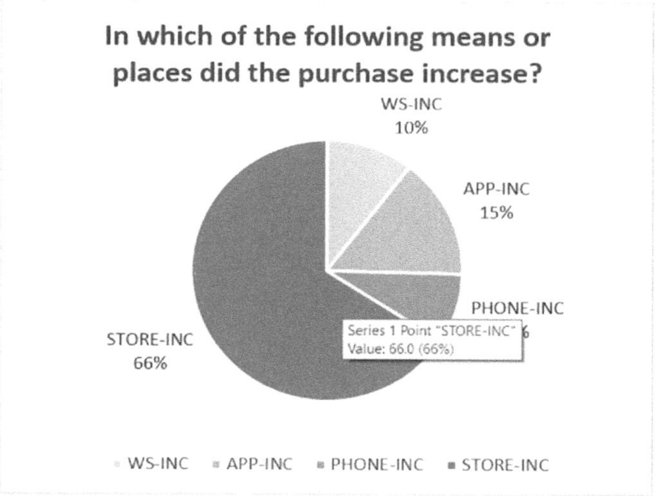

FIGURE 11.4 Where did you buy your groceries?

Source: Elaborated by the author.

11.4.1 CHANGES IN PURCHASE FREQUENCY AND SPENDING

Although the situation of instability and fear due to the new economic situation, spending increases in the new conditions, whether in life or due to new self-protection considerations. In this, as in other crises, we can expect that the mass consumer products industry and mainly the food industry will be little affected, which in the short term continues to be evidenced with substantial increases in the sales of mass consumer products (Nielsen, 2020). The following chart bar in Figure 11.5 shows the income frequency.

Most people earn $242–726 per month (44%), followed by people earning $726–1,210 per month (22.6%), 15.9% earn $1,210–1,694 per month.

Two questions are related to the frequency of purchase and spending:

- Has the number of products in the supermarket purchases increased during the contingency period?
- Has the frequency of purchase of market products increased during the contingency time?

The answers are depicted in Figure 11.6.

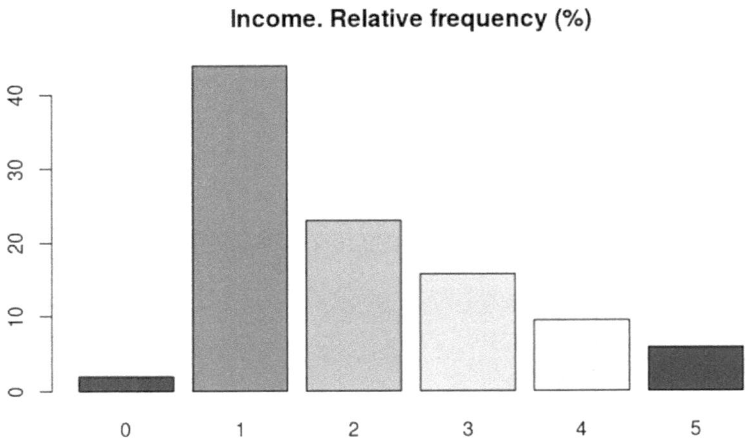

FIGURE 11.5 Income.

Source: Elaborated by the author.

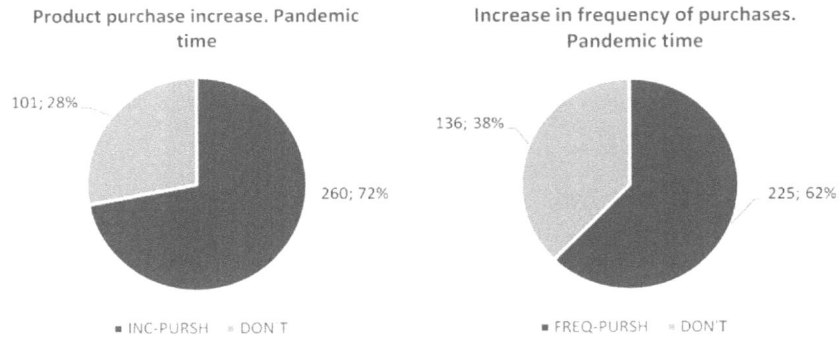

FIGURE 11.6 Increase purchase and frequency.

Source: Elaborated by the author.

Around 72% affirmed that they increased the purchase of products during the pandemic, and 28% affirmed not to have increased. 62% of those surveyed increased their purchase frequency, and 38% did not (Figure 11.7).

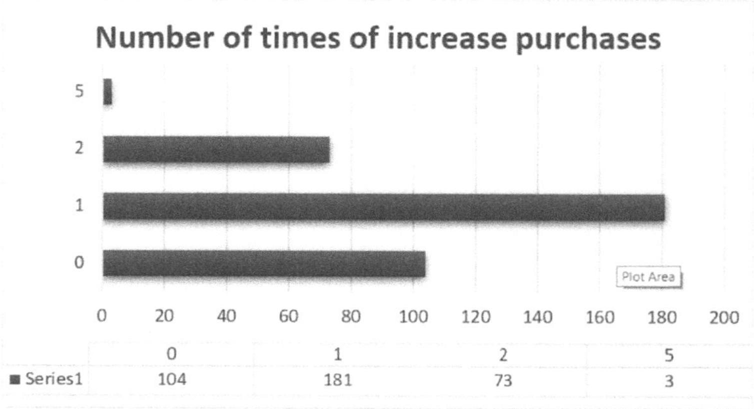

FIGURE 11.7 Increase of the number of times.

Source: Elaborated by the author.

The number of times people went grocery shopping was measured. Due to the chaos resulting from the sudden lifestyle change, consumers began to buy primary products; for example, consumers bought toilet paper in large quantities. Uncertainty and fears of shortages and unusual behavior in consumers are of study interest (Laato, Islam, and Dhir, 2020).

11.5 CONCLUSIONS

Traditional food and grocery stores have been the preferred channel for Colombians to stock up; 32% of households in Colombia were supplied by grocery and convenience retail, supermarket, Superetes, and Discounters during the first quarter of 2019, presenting an increase of 2% points vs. the same period of the previous year (Nielsen, 2019). In addition to the traditional channel, another channel growing is Discounter; Colombians have understood the format and the benefits. This format is relatively recent, the first discounters were opened between 2010 and 2013, and since then, an accelerated spread has begun. Its growth and reception among consumers are surprising in Colombia since it is a relatively new format. The explanation is that these establishments offer substantial discounts on the prices of many essential products compared to traditional stores (Semana, 2017). According to the Colombian E-commerce observatory, the digital operations had a total value of $51.2 billion (COP), of which 30% correspond to product sales and 70% tax payments, public, and private services. Therefore, it is possible that with the new conditions, consumers will change their channel preferences (Dinero, 2019). The following questions were asked to find out if there was a substantial change in channel preference.

Before the pandemic, many brands were in offline channels or perhaps in the multichannel situation, but brands are accelerating the step by having digital channels. COVID-19 has already had a significant impact on consumers and brands. Work and activities from home, changes in the types of products demanded, an insufficient supply of products, and difficulties in supply have forced companies to develop new strategies for the collection and acquisition of products already consumers to adapt to these changes (Knowles, Ettenson, Lynch, and Dollens, 2020). Perhaps in Latin America, the omnichannel culture is not developed for retailers and consumers. Omnichannel requires human and technological resources. The changes are driven by new technologies, such as mobile devices, applications, payment gateways, electronic payments, digital brochures, location applications (Piotrowicz and Cuthbertson, 2014). Due to the overwhelming changes in lifestyles, consumers have had to learn to navigate between consumer channels.

Some changes have been occurring gradually, such as changes in consumption habits, life routines. However, other behaviors have advanced rapidly. Currently, the increase in e-commerce purchases has growth figures much higher than at other times. The country with the most buyers in Colombia, with 12% surpassing Argentina as the electronic commerce leader in Latam. Most of its growth came from people already buying through this channel, but there were also a large number of new users, suggesting that there may now be a lower barrier to entry (Kantar, 2020). However, Colombians continue to prefer the traditional channel and the physical store; organizations must necessarily adopt virtual channels to commercialize their products. There is still a long way to learn about the new consumption habits from the pandemic, so organizations and researchers must continue to monitor consumer behavior changes.

KEYWORDS

- **branding**
- **consumer behavior**
- **multichannel**
- **omnichannel**
- **online shopping**
- **retail strategies**

REFERENCES

Ailawadi, K. L., & Farris. P. W., (2017). Managing multi-and omnichannel distribution: Metrics and research directions. *Journal of Retailing, 93*(1), 120–135.

Ailawadi, K. L., & Keller. K. L., (2004). Understanding retail branding: Conceptual insights and research priorities. *Journal of Retailing, 80*(4), 331–342.

Akter, S., Hossain, M., Lu, S., Aditya, S., Hossain, T., & Kattiyapornpong. U., (2019). Does service quality perception in omnichannel retailing matter? A systematic review and agenda for future research. In: Piotrowicz, W., (ed.), *Exploring Omnichannel Retailing*. https://doi.org/10.1007/978-3-319-98273-1_4. Springer, Cham.

Aronczyk, M., (2020). Brands and the pandemic: A cautionary tale. *Social Media+ Society, 6*(3), 2056305120948236.

Arora, N., Charm, T., Grimmelt, A., Ortega, M., Robinson, K., Sexauer, C., & Yamakawa. N., (2020). *A Global View of How Consumer Behavior is Changing Amid COVID 19*. URL http://tinyurl.com/y83vtejl (accessed on 18 February 2022).

Babu, G. C., & Kumar. S., (2020). A study on consumer behavior pattern shift in the period of corona (COVID-19) pandemic. *Aut. Research Journal,* 523–532.

Barwitz, N., & Maas. P., (2018). Understanding the omnichannel customer journey: Determinants of interaction choice. *Journal of Interactive Marketing, 43*, 116–133.

Blom, A., Lange, F., & Hess, Jr. R. L., (2017). Omnichannel-based promotions' effects on purchase behavior and brand image. *Journal of Retailing and Consumer Services, 39*, 286–295.

Chen, Y., Cheung, C. M., & Tan. C. W., (2018). Omnichannel business research: Opportunities and challenges. *Decision Support Systems.*

Cook, G., (2014). Customer experience in the omnichannel world and the challenges and opportunities this presents. *Journal of Direct Data and Digital Marketing Practice, 15*(4), 262–266.

Dinero. (2019). Retrieved from: https://www.dinero.com/tecnologia/articulo/asi-avanza-el-comercio-electronico-en-colombia/275169 (accessed on 18 February 2022).

Donthu, N., & Gustafsson. A., (2020). Effects of COVID-19 on business and research. *Journal of Business Research, 117*, 284.

Forbes, (2020). Retrieved from: https://www.forbes.com/sites/bradadgate/2020/08/21/nielsen-how-the-pandemic-changed-at-home-media-consumption/?sh=785f44bf5a28 (accessed on 18 February 2022).

Kantar, (2020). https://www.kantarworldpanel.com/co/Noticias/10-millones-de-latinos-compraron-FMCG-en-Ecommerce (accessed on 18 February 2022).

Knowles, J., Ettenson, R., Lynch, P., & Dollens. J., (2020). Growth opportunities for brands during the COVID-19 crisis. *MIT Sloan Management Review, 61*(4), 2–6.

Laato, S., Islam, A. N., Farooq, A., & Dhir, A. (2020). Unusual purchasing behavior during the early stages of the COVID-19 pandemic: The stimulus-organism-response approach. *Journal of Retailing and Consumer Services, 57*, 102224.

Neslin, S., Kenshuk, J., Anand, B., Eric, T. B., John, A. D., Sonja, G., Leonard, L., et al., (2014). The interrelationships between brand and channel choice. *Marketing Letters, 3*, 319–330. http://nrs.harvard.edu/urn-3:HUL.InstRepos:14485590 (accessed on 18 February 2022).

Nielsen, (2019). Retrieved from: https://www.nielsen.com/co/es/insights/article/2019/tendencias-del-consumo-en-colombia-marzo-2019/ (accessed on 18 February 2022).

Nielsen, (2020). Retrieved from: https://nielseniq.com/global/en/insights/analysis/2020/scenarios-beyond-covid-19-rebound-reboot-reinvent-2/ (accessed on 15 March 2022).

Piotrowicz, W., & Cuthbertson. R., (2014). Introduction to the special issue information technology in retail: Toward omnichannel retailing. *International Journal of Electronic Commerce, 18*(4), 5–16.

Saghiri, S., Wilding, R., Mena, C., & Bourlakis. M., (2017). Toward a three-dimensional framework for omnichannel. *Journal of Business Research, 77*, 53–67.

Semana, (2017). Retrieved from: https://www.semana.com/economia/articulo/tiendas-de-descuento-en-colombia/512954/ (accessed on 18 February 2022).

Verhoef, P. C., Kannan, P. K., & Inman. J. J., (2015). From multi-channel retailing to omni-channel retailing. *Journal of Retailing, 91*(2), 174–181.

Von, B. F., (2018). The future of omnichannel retail: A four-stage Delphi study. *Technological Forecasting and Social Change, 132*, 217–229.

CHAPTER 12

Strategies for Nonprofit Organizations in Contingency Environments

CECILIA ISABEL CALDERÓN-VALENCIA

Master of Management and Marketing, UPAEP University, Mexico, E-mail: dra.chechi@gmail.com

ABSTRACT

The goal of this chapter is to present how past and present crises (predictable and unpredictable) have had negative but also positive impacts on the NPO's (nonprofit organization) agenda. In addition, some of the strategies that allowed NPOs to adapt to the conditions and difficulties are presented (innovate instead of costs reduction, increase the connections, increasing transparency), that in the end, could help so that NPOs, who are currently dealing with the difficulties of the COVID-19 environment, can find opportunities, which will allow them to continue with their mission until the contingency passes.

12.1 INTRODUCTION

The NPOs have been affected in the last decades by constant changes such as environmental challenges and economic crises, technological, and digital advances, and changes in demographics and social relationships (Miller, 2018); however, never before had to deal with the physical distancing of individuals as a result of a global health crisis such as we lived in 2020.

Marketing by Contingency in the Time of COVID-19: Overcoming Business Crises and Meeting Marketing Challenges. Alicia De La Peña, PhD & Juan Bernardo Amezcua Nunez, PhD (Eds.)

The results could be anticipated in known contingencies, and organizations can help solve crises through voluntary work and emergency management. However, NPOs are exceptionally vulnerable and at-risk because it is difficult to anticipate and respond to unknown contingencies, even more so if it may lead to physical-social distancing, which means their inability to participate. Physical, social distancing has effects on NPOs because difficult to fulfill their mission, both in the management of human resources (HRs) (volunteers and employees), as well as in the delivery of goods and services produced for beneficiaries (Beckham, Dreher, and Scheck, 2021).

As an example of the effect of the pandemic, it can be stressed that although the new federal legislation that US Congress approved in 2020 (Coronavirus Aid, Relief, and Economic Security Act [ARES Act.]), granted emergency support in the form of loans for small to midsized Nonprofits (500 Employees or fewer), and larger entities (with 500–10,000 employees), and also benefits for donors with a one-time, above-the-line deduction for cash contributions (National Council of Nonprofits, 2020); the Johns Hopkins Center for Civil Society Studies, through their Nonprofit Economic Data Project, published the Impact of the COVID-19 Crisis on Nonprofit Employment in the USA report, which estimates that approximately 1.6 million jobs were lost from March through May 2020. And nonprofit jobs remained 8% down overall all the way through the end of August 2020. The sectors reported as the most affected were Art, Entertainment, and Recreation, Social Assistance, and Educational Services. The least critical were professional and technical services and healthcare (Johns Hopkins Center for Civil Society, 2020).

Further, the revenues of 83% of the NPOs in the USA have suffered a fall as a result of the global COVID-19 pandemic, canceling events, reducing donations, and fewer grants. Besides, NPOs that provide food, shelter, or support for medical care report an increase in the demand for their services (Beckham, Dreher, and Scheck, 2021).

As the contingency approach suggests, the performance of organizations depends on how their characteristics (the organizational structure, the size of the organization, the gender diversity, and the organizational strategies) adapt to the changing situations of the environment (Mohammad and Bujang, 2019; Park, 2020). This flexibility allows the organization to cope in a constantly changing environment; therefore, organizations adjust to contingencies because they need to adapt to avoid losses in their performance (Donaldson, 2001).

The competitive environment forces NPOs to adopt the concepts and strategies of for-profit organizations (Mikołajczak, 2019), even if the purpose is different since NPOs do not seek results aligned to profits nor to obey investor's interests, but who seek to invest in people to address the social concerns (Kong, 2008). This chapter seeks to describe several of the strategies employed by NPOs in past crises, and in the current COVID-19 environment, in order to offer strategic alternatives to deal with the current and future crises.

12.2 THE NPOS TYPOLOGY

To understand the impact of physical distancing, as a critical event for NPOs, its operation and structure (members such as board, administrative management, contracted employees, volunteers, foundations, corporate philanthropies, and donors) must be understood, which allows it to access the necessary resources to achieve its purpose of attending a group of beneficiaries (Cavazos-Arroyo and Calderón-Valencia, 2019).

NPOs can be classified in different ways according to the context in which they are developed and the legal framework in which they operate. It can be observed, in the United States, 71.9% of NPOs correspond to Public Charities, while Private Foundations represent 6.7% of organizations (Golensky and Hager, 2020).

Analytical-theoretical typologies include variables/dimensions or categories (e.g., example size, internal political structure, dependence on external control or resources, societal functions, sources of support, location, the classes and characteristics of their members, intimacy among members, main dependence on one or another kind of incentives – material, solidary, or purposive-, and main type of beneficiary) (van Deth, Edwards, B˘adescu, Moldavanova, and Woolcock, 2016; p. 183).

Another typology of NPOs is based on their purposive activity, just as voluntary organizations and associations. "Nonprofit agencies operated mainly by paid staff (e.g., a nonprofit hospital, museum, or university) and voluntary associations operated mainly and often solely by volunteers" (Smith et al., 2017; p. 93).

NPOs tend to be normative organizations and can be classified by size and activity type, such as formal and informal activities. However, all of them require resources to operate. Most of the contributions come from public and philanthropic funding (government, UN, UNICEF) as well as from large corporate donors (Carlos Slim Foundation in Mexico) and individual donors (who collect economic resources, food, consumer goods, medicines, and even organs and blood).

In the social economy field, and from the de organizational approach, NPOs are recognized as primarily trading organizations and are considered part of social enterprises, including cooperatives, mutuals, community business, and voluntary/Nonprofit organizations (Spear, 2006; Doherty, 2013).

Also, NPOs are known as part of the organized civil society sector; for example, grassroots associations (GA) is defined as "nonhierarchical, social justice-focused, community-based associations of activists and residents who come together over an extended period of time to address a local social problem" (Nickels and Clark, 2019, p. 201). Therefore, it can be said that there is a wide variety of NPOs, with different formats and different ways of operating.

12.3 THE EFFECT OF THE PAST CRISES IN THE NONPROFIT SECTOR AND NPOS PERFORMANCE

Although normally crises have a specific origin, such as economic crises, environmental crises, humanitarian crises related to wars and immigration, political crises, etc. (see Table 12.1), all of them affect the countries' economy, and therefore for organizations. However, the health crisis caused by pandemics is unprecedented. Since it forces people to keep their distance to take care of their health; this fact has a significant impact on all sectors, both from the supply side (productivity reduction due to the decline of working hours and/or minimum number of employees in a single physical space) and from the consumption side (reduction of consumption in public places for fear of contagion.

Many studies stressed that in the years of economic expansion, the civil sector prospered as well, and in crisis time, the NOPs can suffer financial problems, but some of them can be resilient (Horvath, Brandtner, and Powell, 2018). Table 12.1 has been extracted by publications along the time, showing different examples of the effect of several different crises in the nonprofit sector, and the strategies adopted by them.

TABLE 12.1 The Effect of Crises in NPOs

Source	Country	Year and Type of Crisis	Negative Impact	Positive Impact	Adaptation Strategy
Pospíšil (2013)	Czech Republic	2008–2012 global recession	Briefly decreased funding from the state.	Private donors (individuals and businesses) have been largely unaffected by the crisis.	To improve the relations with the public and the corporate world, to attract more citizens and companies to nonprofit activities, to win more participation and support from them, and to increase civic engagement levels. Demonstrate how much they can do for society in hard times, boosting thus their self-confidence as well as their image in the eyes of the public.
Kákai (2014)	Hungary	2008–2010 Global economic crisis	Civic presence and activity decreased. The lack of legal regulations and sanctions to stop the operations may result in unclear and illegal operations from NPO representatives.	The citizens react to the world by embracing them through getting self-organized in order to achieve smaller or bigger community objectives. The weight of the nonprofit sector grew during the years examined.	In NPO with scarce resources depends mainly on its leaders' skills and creativity. In organizations with substantial resources, increase the connections (through contracts and subsidies) with the state and local governments or their institutions

TABLE 12.1 *(Continued)*

Source	Country	Year and Type of Crisis	Negative Impact	Positive Impact	Adaptation Strategy
Hu and Kapucu (2017)	USA–Central Florida	2008 financial crisis	Aggravated financial conditions. On average, NPOs decreased their programs and fund-raising expenditures. Decrease of total revenue and total assets.	Organizations that are more confident in their revenue sources may continue to provide their services. Large NPOs are more likely to spend more money to sustain existing programs.	Strategic plan implementation and board involvement. Adding new programs, suspend existing programs or staff. Building and expanding programs through collaboration.
Tzifakis, Petropoulos, Huliaras (2017)	Greece	2015 Global economic crisis	A sharp decline in revenues. Dependence on external founding	Many NPOs instituted reforms that increased efficiency, reaching record level numbers of volunteers, and an impressive rise in funding by the private philanthropic foundation	Decrease their operating costs mainly in expenses. Restructure operations to reach efficiency rather than decrease its activities. Increase the response to their donor's expectations in order to diversify their sources of income

TABLE 12.1 *(Continued)*

Source	Country	Year and Type of Crisis	Negative Impact	Positive Impact	Adaptation Strategy
Simsa (2017)	Austria	Autumn and winter of 2015- refugee crisis	Changing conditions and a lack of information. Challenges regarding the late and insecure funding of activities, information deficits, the fast organizational growth, and the coordination of high numbers of volunteers. Lack of official, governmentally organized activities are not only legally problematic but also generates physical and psychological stress experienced by volunteers and financial stress of NPOs.	NPOs experienced substantial growth and gained legitimacy by high public attention and positive response in media and social media. Planning and coordination of voluntary work. Communication between public bodies and civil society actors improved significantly.	Especially established organizations are used to select and train their voluntary staff and usually work on the basis of clear formal structures. Psychological care and support for volunteers organized by civil society.

TABLE 12.1 *(Continued)*

Source	Country	Year and Type of Crisis	Negative Impact	Positive Impact	Adaptation Strategy
Nickels and Clark (2019)	USA	The Flint Water Crisis, 2014	Focus on the needs of the elite business community and the importance of wealthy philanthropists over the people. The nonprofit sector has done very little to promote social justice and amelioration in the crisis; on the contrary, it has promoted activists who focus on the problem causes and try to find the culprit to pay for their crime.	NPOs offer different yet important skill sets and resources to a community.	NPOs would benefit from recognizing their embeddedness within the local power structure and that ties to other "benevolent donors," which have important ramifications (networks) on their crisis response.

Note: Original based on Pospíšil (2013); Kákai (2014); Hu and Kapucu (2017); Tzifakis, Petropoulos, and Huliaras (2017); Simsa (2017); Nickels and Clark (2019).

As organizations lean towards adopting new formats that allow them to reach many more people, they can help meet their needs to a greater extent than they could in better economic times. Some NPOs are more focused on continuing their operations without focusing on specific goals and programs (Beckham, Dreher, and Scheck., 2021); however, it is mandatory to identify how they can achieve it; the strategies listed below were extracted from the literature presented, and from the interviews reported.

12.4 THE STRATEGIES DEALING WITH THE COVID-19 CRISIS

While facing financial uncertainty, nonprofit organizations may use several strategies such as adding new programs, implementing or expanding joint collaborative programs, increasing their earned income, and starting or increasing their involvement in advocacy. Also can apply reactive strategies such as cutting off existing programs, reduce the staff, eliminating volunteer work (Hu and Kapucu, 2017), or activating alarms to protest against the origin of the crisis (Nickels and Clark, 2019).

Based on interviews with executives of three NPOs, it can be noticed that the phenomenon of the pandemic considerably reduced the action of donors and volunteers, and therefore access to resources. For example, one of the interviewees specified:

"*For the well-being and health of the volunteers, the board members decided to avoid and suspend contact with volunteers, thanking them for their participation and urging them to continue being attached to the NPO to join in the future*."

The second interviewee reported that NPO decided to minimize the hired staff and retain only the services of employees who act from a distance, only as contacts or facilitators to reach the beneficiaries, but without having any physical or direct contact with them.

On the contrary, in the third case, the NPO decided to increase operations, incorporating online activities, inviting the community of volunteers to participate by offering the beneficiaries some professional services (medical and psychological services). Besides, a fundraising event was held and was presented for the first time in a streaming video format, allowing them to get closer to their donor organizations and obtaining encouraging results.

NPOs that are more limited in access to financial resources "may divert their attention from mission accomplishment to coping with financial pressures." (Zietlow et al., 2018; p. 29), and the most limited NPOs are the least formal, and that is why it is so important that NPOs formalize their activities through a legal and organizational structure that allows them to access more sources of resources (Simsa, 2017).

Donations also suffered a significant decrease in all three cases (around 48% between March and September 2020), which made it hard, above all for less formal organizations, to continue with their activities to help vulnerable groups such as women and the elderly. However, two of the NPOs had to migrate the provision of their services from a face-to-face format to a virtual format, enabling them to continue with the provision of services, primarily on health issues and the provision of technical services.

From all the information gathered, it is possible to describe the strategies listed below for dealing with critical environments.

12.4.1 INNOVATE INSTEAD OF COSTS REDUCTION

Organizations that are larger and more formal tend to reduce costs; however, those that are smaller and have no structure but a volunteer-based organization, without a formal executive director, may not reduce costs because that would imply ceasing to operate (Kákai, 2014). Cost reduction can occur through cuts in personnel and sacrificing some scarce beneficial programs; however, the strategy applies only to organizations that have access to constant sources of funding, which does not happen in most NPOs in high-uncertainty environments.

In contrast, innovation can be achieved by reshaping business models in response to the crisis, including iterative innovation based on trial and error (Reficco, Layrisse, and Barrios, 2020). In this more than in other crises, the use of information and communication technologies has been crucial for NPOs to continue operating. However, this process of technology adoption must be quick and understandable for all members of the organization; in this way, NPOs can formalize the activities by hiring an executive director with the skills and technical knowledge to embrace the transformations and to use technology to encourage the mission of the NPO, can be part of the business model innovation to cope with crises.

12.4.2 INCREASE THE CONNECTIONS

Through the connections of the board of directors and executive director, more, and better resources from public and private donors may be accessed; therefore, the participation of these actors in different socializing activities making the mission of the NPO known is extremely important. NPOs can be an engine of stability in economic crises through maintaining their activity "despite zero or negative rate of return diversifying funding sources and entrepreneurship, increasing cooperation with other sectors (governmental and private)" (Dominowska, 2016; p. 179).

Especially in the NPOs that handle a large number of resources, the connections should be increased through contracts and subsidies from governments and institutions; also, solid relationships between the executive director and the executive board should be fostered to produce more effective organizations. Board commitment and financial donations are highly dependent on donations, and donations are related to financial performance (Hu and Kapucu, 2017). Likewise, donors can be more generous in times of crisis with NPOs causes, but this behavior is temporary, and when the crisis ends, the behavior fades.

12.4.3 INCREASING TRANSPARENCY

Open communication and trusted partnerships between funders and NPOs are capital to identify the ways to support low budgets and goals changed or suspended. "Transparency is the degree of information and the attitude with which the accountability process is performed" (Valencia, Queiruga, and González-Benito, 2015; p. 2522). Transparency is crucial to gain the confidence of stakeholders, and communication mechanisms are strong tools to transmit transparency. Donors appreciate having access to independent, free, objective, and consistent information on NPO operations to help them decide which organization to partner with and to track their donations. No matter the nature of the situation, transparency creates a sense of formality and powerful trust in the community. Finally, NPOs must prioritize the needs of their community and adapt their services to better serve them through reshaping their strategy and operations in contingency environments.

12.5 CONCLUSION

The impact of some crises on NPOs have been well documented in the past had very damaging effects; however, they presented opportunities that allowed them to resist and take advantage of that contingency environment. At the same time, some NPOs, have been able to sustain themselves through the current health contingency, by adopting many strategies that allow them to deal with difficulties and take advantage of favorable opportunities. This chapter takes a realistic view, offering strategies to NPOs in contingency environments.

KEYWORDS

- **costs reduction**
- **COVID-19 crisis**
- **nonprofit sector**
- **NPOs typology**
- **transparency**

REFERENCES

Beckham, H., Dreher, J., & Scheck. D., (2021). Building resiliency in times of crisis. *Stanford Social Innovation Review, 19*(1), (Sp)15.

Cavazos-Arroyo, J., & Calderón. V. C. I., (2019). Two paths to the same destination: Participation levels as a result of affiliation networks in civil society organizations. *Journal of Nonprofit & Public Sector Marketing, 31*(2), 139–163.

Doherty, B., (2013). Social enterprise management. *Social Enterprise: Accountability and Evaluation around the World,* 28.

Dominowska, J., (2016). The role of third sector in economic crises. *Economics and Law, 15*(2), 179–191.

Donaldson, L., (2001). *The Contingency Theory of Organizations*. Sage.

Golensky, M., & Hager. M. A., (2020). *Strategic Leadership and Management in Nonprofit Organizations: Theory and Practice*. Oxford University Press.

Horvath, A., Brandtner, C., & Powell. W. W., (2018). Serve or conserve: Mission, strategy, and multi-level nonprofit change during the great recession. *Voluntas: International Journal of Voluntary and Nonprofit Organizations, 29*(5), 976–993.

Hu, Q., & Kapucu. N., (2017). Can management practices make a difference? Nonprofit organization financial performance during times of economic stress. *Journal of Economics and Financial Analysis, 1*(2), 71–88.

Kákai, L., (2014). Hungarian nonprofit sector twenty years after: The situation of the third sector and impact of the economic crisis. *Politeja: The Journal of the Faculty of International and Political Studies of the Jagiellonian University, 11*(28), 83–113.

Kong, E., (2008). The development of strategic management in the non-profit context: Intellectual capital in social service non-profit organizations. *International Journal of Management Reviews, 10*(3), 281–299.

Mikołajczak, P., (2019). Becoming business-like: The determinants of NGOs' marketization turning into social enterprises in Poland. *Oeconomia Copernicana, 10*(3), 537–559.

Miller, E. W., (2018). Nonprofit strategic management revisited. *Canadian Journal of Nonprofit and Social Economy Research, 9*(2).

Mohammad, H. S., & Bujang. I., (2019). Firms' size as a contingency approach in the relationships of intellectual capital, capital structure and financial performance. *International Journal of Academic Research in Business and Social Sciences, 9*(3).

National Council of Nonprofits, (2020). https://nonprofitquarterly.org/how-nonprofits-can-utilize-the-new-federal-laws-dealing-with-covid-19/ (accessed on 18 February 2022).

Nickels, A. E., & Clark. A. D., (2019). Framing the Flint water crisis: Interrogating local nonprofit sector responses. *Administrative Theory & Praxis, 41*(3), 200–224.

Park, S., (2020). Size matters: Toward a contingency theory of diversity effects on performance. *Public Performance & Management Review, 43*(2), 278–303.

Pospíšil, M., (2013). *Czech Giving in the Times of Economic Crisis*. Masaryk University. Available at: SSRN 2365709.

Reficco, E., Layrisse, F., & Barrios. A., (2020). From donation-based NPO to social enterprise: A journey of transformation through business-model innovation. *Journal of Business Research.*

Simsa, R., (2017). Leaving emergency management in the refugee crisis to civil society? The case of Austria. *Journal of Applied Security Research, 12*(1), 78–95.

Smith, D. H., Stebbins, R. A., Grotz, J., Kumar, P., Nga, J., & Van, P. P., (2017). Typologies of associations and volunteering. In: *The Palgrave Handbook of Volunteering, Civic Participation, and Nonprofit Associations*. Springer.

Spear, R., (2006). "Social entrepreneurship: A different model?" *International Journal of Social Economics, 33*(5, 6), 399–410. https://doi.org/10.1108/03068290610660670.

The Johns Hopkins Economic Data Project. Johns Hopkins Center for Civil Society Studies. http://ccss.jhu.edu/labor-day-2020-coronavirus/ (accessed on 18 February 2022).

Tzifakis, N., Petropoulos, S., & Huliaras. A., (2017). The impact of economic crises on NGOs: The case of Greece. *Voluntas: International Journal of Voluntary and Nonprofit Organizations, 28*(5), 2176–2199.

Valencia, L. A. R., Queiruga, D., & González-Benito. J., (2015). Relationship between transparency and efficiency in the allocation of funds in nongovernmental development organizations. *Voluntas: International Journal of Voluntary and Nonprofit Organizations, 26*(6), 2517–2535.

Van, D., Edwards, Bˇadescu, Moldavanova, & Woolcock (2017). Associations *and Social Capital*. In: *The Palgrave Handbook of Volunteering, Civic Participation, and Nonprofit Associations*. Springer.

Zietlow, J., Hankin, J. A., Seidner, A., & O'Brien, T., (2018). *Financial Management for Nonprofit Organizations: Policies and Practices.* John Wiley & Sons.

Glossary

A

Anthropomorphic Brand Traits – brand description based on human beings and personality traits. It is to make a brand a human-like entity. Also called Brand persona or personification.

B

Big Data – huge quantity of data (structures or not structured, numerical, text, sounds, images) produced by an entity, as a consequence of operation, commercialization, and delivery of products and services, as well as from interactions with stakeholders. Commonly, the collection, organization, integration, and analysis of big data details discovering patterns, trends, associations, and causal effects that enable firms to better decisions taking and better their competitive positions.

Black Swan – a rare and unpredictable market event that might cause severe damages to an economy.

Brand Empathy – is related to an emotional connection between brands and consumers. It is about brands understanding customer needs and putting in their shoes.

Brand Loyalty – positive association developed with a brand, that makes the consumer to look for it, wait for it, no matter other offers of the competitive brands.

Brand Omission/Failure – the inability of a brand to meet customer needs and wants. It is a brand failing to understand what consumer needs at both emotional and rational level.

Brand Perception – thoughts and feelings a consumer has about a brand.

Brand Relationship – the relationship that consumers have, think to have, or feel with a brand. It is the result of repeated interactions between a consumer and a brand.

Brand Transgression – is a violation of the assumed consumer-brand relationship relevant norms.

Brick and Mortar Stores – physical stores in which personal contact or assistance is necessary for purchasing or consuming products or services.

C

Company's Brand Knowledge (CBK) – the degree of employees' knowledge about the company goals, brand values, value proposition, brand identity, and communication.

Consumer Behavior – the study of how individuals process information, make decisions, consume, and dispose of products.

Consumer Trust – an intention or behavior that reflects consumer's willingness to rely on a company despite the presence of a certain degree of uncertainty and vulnerability.

Consumer-Centric Socially Responsible Consumption – a form of socially responsible consumption that considers the impact that consumption has on other consumers.

Consumption Effects Due to Pandemic – these are the drastic changes in consumption that arise due to confinement conditions and economic and health uncertainty.

Contingency – event or circumstance which is possible but cannot be predicted with certainty.

Coopetition – cooperation between rival companies.

Coping Strategies – mechanisms that people use to face stressful situations and maintain emotional well-being.

Corporate Social Responsibility – a company's commitment to operate in a way that improves society.

COVID-19 – infectious disease caused by a recently discovered type of coronavirus.

COVID-19 Lockdown – due to the COVID-19 pandemic, governments state rules for their inhabitants to control the pandemic. Some examples of these rules that may vary according to the country: stay-at-home orders, quarantines, identification of essential jobs, healthy social distance, etc.

COVID-19 Pandemic – it is a pandemic that emerged in early 2020 due to the expansion of the Corona Virus; it received its name because it emerged at the end of 2019 in China.

Covidiot – individual who behaved irresponsibly with regards of CDC's recommendations and guidelines, putting others at risk in terms of spreading the disease.

D

Dedicated Microsite – a temporary website built for a limited time, commonly used for specific marketing campaigns. Usually, microsites have their own domain name, and focused content, helping a brand achieve different marketing goals.

Descriptive Modeling – application of analytics algorithms focused on a description of a variable or a set of variables immersed or related with a studied phenomenon (the current status).

E

e-Commerce – selling and buying products using the Internet.

Employee Net Promoter Score (eNPS) – a metric about how likely the employee will recommend a company to a friend or colleague to work in it.

Engagement – a connection among two entities (i.e., persons, consumers, firms) during a period of time.

G

Germaphobia – phobia produced by the potential or actual effect of contact with microscopic bacteria or viruses, in public places or on objects of daily use (especially on the health of the person).

I

Influencer – an individual who has a distinct voice and is used in the context of social media to denote the ability to persuade their audience to acquire products or services based on their recommendations.

Influencer Marketing – strategic decisions made by an organization/brand that include influencers as part of their selected communication channels in order to appeal to the specific audience of said individual.

In-Homing Behavior – trend centered on purchasing and consuming goods or services in the home's consumer or buyer.

Internal Marketing – the promotion of a company's brand within the organization considers the employees the first customers.

L

Lifestyle – the way in which a person or group of persons live; generally associated with consumption, as an example: healthy lifestyle.

M

Marketing Channels – different organizations that help to put the product or service available for the consumers.

Merchandizing – it is a set of promotional activities that contribute to selling products in a store. It usually includes the creative display of products that are for sale.

Multichannel – it is the marketing separately in physical and virtual sales channels of products and services.

Multiple Channels – the use of brick-and-mortar stores and online channels to satisfy consumers' needs.

N

NPO – non-for-profit organizations.

O

Omnichannel – the integration of customer, brands, and retail channel interactions to create a seamless retail experience.

Online Channels – the use of the Internet (i.e., website), social media to allow online sales and to satisfy consumers' needs.

Organizational Citizenship Behavior (OCB) – the employee's behaviors that are beyond the job description that promotes the effective functioning of the organization.

Organizational Commitment – an employee's psychological attachment to stay working for an organization, measured in three dimensions: affective, normative, and continuance.

P

Panic Buying – involves buying unusually large amounts of a product as a result of the anticipation of a disaster, crisis, price increases or supply shortage.

Perceived Organizational Support (POS) – a measure of the employee's beliefs how an organization values his/her contribution and cares about his/her well-being.

Personified Brands – brands that are entitled human being and personality traits. Is giving a brand a human face to create enriching stories that help connect to consumers.

Phenomenological Approach – research based on a qualitative method with in-depth interviews, developed with the specific goal of learning about the experience of the interviewed through the detailed description of the event and its context.

Predictive Modeling – application of analytics algorithms focused on unveiling relationships or grouping/classification structures among a set of variables immersed or related with a studied phenomenon. In this way, the model will produce a new predicted value or classification/categorization of a target variable in a research setting.

Prescriptive Modeling – application of analytics algorithms focused on taking advantage of descriptive and predictive models for suggesting or recommending options or actions to develop or implement based on a set of variables immersed or related with a studied phenomenon. Commonly, prescriptive modeling involves an auto-learning process for model improvement and optimization for decision making (a self-learning process). AI (Artificial Intelligence) is one of the widest prescriptive models used nowadays.

Psychological Contract – an implicit agreement between the brand and the consumer, where expectations for the relationship are established.

Purposeful Sampling – sampling technique in which the interviewed participants are selected based on the judgment of the researcher considering the profile representative of the population under study.

R

Retail – the sale of goods in small quantities for the final consumer.

Retail Brands – they are brands that distribute products to the final consumer and sell products from other brands under their own.

Retailment – a brand strategy designed to connect with the customers in a store. It comes from the union of two words: retail and entertainment. Therefore, the goal of retailment is to provide a different shopping experience to clients, usually with the aid of technology, like touch screens, novel products' layouts, and sampling of products.

S

Safe Distancing Strategies – strategies employed at stores and restaurants to help customers and employees keep a safe space between each other to reduce the spread of COVID-19.

Social Identity Theory – social identity is a person's sense of who they are based on their group membership(s), as proposed by Henri Tajfel (1979).

Social Influence – ways in which people modify their behavior to meet the expectations of the social environment.

Social Media – communication channels that allow multiple audiovisual material to be shared with a specific audience.

Subscription Boxes – business model and marketing strategy consisting of the delivery of particular specialty products, packaged together to allow consumers to experiment with new brands every few weeks.

T

Traditional Stores – it is the oldest retail distribution channel in Colombia; it can be related to grocery stores and convenience stores. Technically it receives the name of TAT.

U

Unboxing Videos – videos created either by influencers or individuals, with the goal to show the content of a package. Creators of unboxing videos capture the process of unpacking a product, ad a detailed explanation, and even some music. Once the video is recorded, they share the video with others using YouTube, TikTok, or Instagram.

V

Virtual Channels – these are all channels for the sale, promotion, and distribution of products through digital media, platforms, and applications.

Z

Zoom Workshops – remote workshops that can be used by individuals or businesses to capture an audience and deliver helpful content. As a marketing tool, brands are using virtual workshops to keep customers engaged and interested in the brand while providing instructions to create different items: food dishes, artistic items, etc.

Index

G

H

I

P

T